D1603092

MT 333.95 SMITH
Smith, Bruce L.,
Wildlife on the wind :

DATE DUE

FEB 1 4			

Wildlife on the Wind

A Field Biologist's Journey and An Indian Reservation's Renewal

Bruce L. Smith

Utah State University Press
Logan, Utah
2010

Copyright © 2010 Bruce L. Smith
All rights reserved
Utah State University Press
Logan, Utah 84322
www.USUPress.org
Manufactured in the United States of America
Printed on acid-free, recycled paper

978-0-87421-791-9 cloth
978-0-87421-804-6 paper
978-0-87421-792-6 e-book

Library of Congress Cataloging-in-Publication Data

Smith, Bruce L., 1948-
Wildlife on the wind : a field biologist's journey and an Indian reservation's renewal / Bruce L. Smith.
  p. cm.
Includes bibliographical references and index.
 ISBN 978-0-87421-791-9 (cloth : acid-free paper) – ISBN 978-0-87421-804-6 (pbk. : acid-free paper) – ISBN 978-0-87421-792-6 (e-book)
 1. Shoshoni Indians–Ethnobiology–Wyoming–Wind River Indian Reservation. 2. Arapaho Indians–Ethnobiology–Wyoming–Wind River Indian Reservation. 3. Wildlife management–Wyoming–Wind River Indian Reservation. 4. Biology–Fieldwork–Wyoming–Wind River Indian Reservation. 5. Wind River Indian Reservation (Wyo.) 6. Smith, Bruce L., 1948- I. Title.
 E99.S4S65 2010
 333.95'4160978763–dc22
                                    2010024664

*To the people of the
Eastern Shoshone and Northern Arapaho nations,
past, present, and future.*

But now, says the Once-ler, now that you're here, the word of the Lorax seems perfectly clear. UNLESS someone like you cares a whole awful lot, nothing is going to get better. It's not.

—Dr. Seuss, *The Lorax*

Contents

Introduction   *ix*

Part I
Chapter 1: Gettin' There   3
Chapter 2: On the Reservation   23

Part II
Chapter 3: First Elk   49
Chapter 4: Mountains and Sky   75
Chapter 5: Stranded   96
Chapter 6: The Way It Was   120
Chapter 7: Younger Kids   143

Part III
Chapter 8: On the Same Page   159
Chapter 9: Game Code   176
Chapter 10: Upshot   193

Epilogue   *206*
Acknowledgments   *211*
References   *215*
Index   *220*

Illustrations

Dr. Frank Enos	16
Shoshone Chief Washakie	25
Map of Wind River Indian Reservation	34
Biological technicians clipping vegetation	40
Elk survey from a Cessna	51
Roberts Mountain	57
Crowheart Butte	62
Black bear in Bull Lake Canyon	80
Elk cows and calves in Bull Lake Canyon	84
Bull moose	90
On Trail Ridge glassing for bighorn sheep	97
Hiking down Trail Ridge from disabled helicopter	110
Herman and Rachael Lajeunesse	126
Bighorn sheep in the Wind River Mountains	129
Bighorn winter range	132
Field trip with Joint Business Council	161
American kestrel	166
Pronghorn antelope tangled in fence	172

Introduction

There is nothing quite like it, the anticipation I feel buckling into the safety harness of the bolt-upright seat of a helicopter, hearing the turbine fire up, feeling the chassis shudder, and seeing the drooping rotors spin to life and flatten overhead. With the mountains newly frosted, this promises to be a glorious day. Creeping down from the Wind River Range's frozen summits, salmon pink light awakens the dense bands of evergreen forest, stabs the small parks tucked within, and stirs horned larks into whirling, tight flocks against foothills of gray-green sage. With snow crystals sweeping the acrylic bubble—the frigid air unable to retain a drop more moisture—the chopper lifts me up Dinwoody Ridge.

On Dinwoody's ancient cliffs, early Americans worked stone on patina, bequeathing us etched canvases of these original "rock stars." Where they hunted bighorns with arrows fashioned of willow shafts, obsidian heads, and duck quill rudders, I now hunt descendant sheep, elk, mule deer, and moose. My weapons, however, are only my eyes, a tape recorder, and a lapful of topographic maps with which to capture each enduring detail.

🦌 🦌 🦌 🦌 🦌

Long before I counted bighorn sheep from helicopters above Wyoming cliffs engraved with ancient petroglyphs, before I became a professional wildlife biologist, I was a nascent naturalist. My adolescent curiosity in wild things and the complexity of Nature drew me to a conservation career. Far from western Michigan's checkerboard farmlands and mixed hardwoods, the sprawling basins and towering peaks of the Rockies called to me. I would live among the highest-dwelling of North America's big game, and eventually work with every large mammal species in the western U.S. At an early stop in my 30-year career as a federal wildlife biologist, I became a conservation partner with two Plains Indian tribes.

Interest in American Indians has swelled in recent years. Hundreds of books cover every conceivable topic—history, cultures, traditions, sociology, reservation economics, and natural resources. Even the political climate toward Native Americans has changed with the 2008 presidential primary candidates visiting reservations and President Barack Obama recognizing them among other ethnic groups in the inclusive message of his speeches. Many others have worked with American Indians to preserve or restore their cultural heritage, including the wildlife that suffuses their traditional lifestyle and religion. In fact, native people and wildlife suffered a parallel fate as the continental tsunami of Euro-American advance decimated the great herds of buffalo (American bison), elk, pronghorn antelope, bighorn sheep, grizzly bears, and wolves. In the aftermath, this is the first book written about wildlife restoration on any of the 310 Indian reservations in the contiguous U.S.

Wildlife biology and management are unlike most fields of natural science. The focus is the study of beautiful and charismatic, sometimes threatening and creepy living things. They beguile us with their uniqueness and unpredictability. Because wildlife is a product of the land and water, to understand the animals we must understand where and how they live. By extension, "wildlifers" (those who study or manage wildlife) must grasp a host of disciplines, including geology, soil science, plant ecology, chemistry, evolution, physiology, and animal behavior. Without the additional knowledge of environmental influences that affect animal populations and communities—farming, pollution, weather, fire, livestock and forestry management, etc.—wildlifers can no more effectively do their jobs than can physicians unfamiliar with all body parts, their interrelations, and responses to the environment. In other words, wildlife science is an integrative profession. In part, that's what attracted me, the opportunity and challenge to understand the "big picture."

What I didn't learn or fully appreciate in college were the politics and public relations demands of a resource management career. Wildlife management is mostly people management. Animals do quite well on their own—if we allow them. Given humanity's now ubiquitous footprint on the biosphere, wildlife management and conservation have become increasingly complex and daunting enterprises. One indication comes from a survey of 400 eminent biologists by the American Museum of Natural History. The results suggest that the current rate of species extinction—30,000 annually—is the fastest since the last mass extinction 65 million years ago. That ancient apocalypse at the close of the Cretaceous Period wiped out half of all species, famously including the dinosaurs.[1]

But to rescue species was not why I signed on. My interest was to sustain whole communities of species. After honing my skills as a budding wildlifer during stints in Montana, Wyoming, Washington, and California, I landed what proved to be the job of a lifetime. I became the first wildlife biologist to work on Wyoming's Wind River Indian Reservation. This book details my wildlife restoration efforts with the Northern Arapaho and Eastern Shoshone nations. As a non-Indian among these people, I recount our cooperative endeavors to change the course of conservation for America's most charismatic wildlife.

I liken the challenge to landing in conservation's Garden of Eden. Lacking any previous records of the reservation's wildlife, I embarked on a four-year effort to learn all I could about those 2.2 million acres' imperiled big game. During days afield on foot and horseback and hours spent surveying pronghorns, bighorns, moose, and elk from the air, I describe the adventures and close calls of a biologist's fieldwork. Whether stalking newborn elk calves, tending an injured falcon, sidestepping bears, or number-crunching habitat data, I found each day rewarding. Field studies, spent typewriter ribbons, classroom time with Indian children, hammering out hunting regulations with tribal councils, and tense public meetings all were essential ingredients of the same recipe—to restore depleted game populations for two Plains Indian nations.

Beyond the struggling wildlife, I was humbled and enriched to work with descendants of people who predated my Dutch and Lithuanian ancestors' New World arrival by 15,000 to 20,000 years. I had never met a Native American—at least not that I knew of—before joining the Marines. A decade later I knew scores of them. Moreover, I shared a dream of reweaving their cultural fabric with the wild creatures the Creator had given them. Toward our common purpose, I gained the trust of the Shoshone and Arapaho people to forge an inspiring partnership.

This book relates the rewards of sustaining life in wild places, as well as my firsthand education about the cultural significance of wildlife to native people. Even more, it is about the triumph of perseverance and shared commitment to pass on to future generations the inheritance of Nature with which we're all blessed. The conservation success at Wind River serves as a shining model for other Indian nations seeking to preserve their own wildlife heritage.

PART I

Only to the white man was nature a "wilderness" and only to him was the land "infested" with "wild" animals and "savage" people. To us it was tame. Earth was bountiful and we were surrounded with the blessings of the Great Mystery.

—Chief Luther Standing Bear, Oglala Sioux

1

Gettin' There

What's it like working with Indians? Early on, the question unsettled me. Not because I had no consequential answer, but because of how it struck my sensibilities. Often I interpreted the question's tone to probe dark secrets about the Indian people, rather than to learn about our shared endeavors. Perhaps I would confirm someone's preconceived notions or disclose a shocking revelation. I heard the question directed toward those who were different from us, as if inquiring about strange aliens.

Ironic this seems to me now for the obvious reason. If anyone, we non-Indians are the aliens. But there's another paradox. For years I was among those who were uninformed about who these people living in *our* country were. Native Americans—the politically correct term invented by us white folks decades ago—rings hollow with a Sioux or an Iroquois or a member of any of 500 other tribes scattered from Plymouth Rock to California's Mission Valley. At least that was true three decades ago when I was among native people. Don't get me wrong. I understand our genteel intent. But to the ears of those whose homeland was overrun and renamed America, it's a subtle reminder of Euro-Americans' past treachery.

As it turns out, "Indians" is the label that North America's indigenous people prefer. At least that's the case when speaking inclusively of the continent's first people. But assuming that all Indian nations are equivalent to one another is like saying that the British are like the French. Cherokee or Apache, their tribal affiliation is how most, at their core, still think of themselves.

So what's it like working with Indians? Better yet, what will it be like? That was my own question in 1978. As a transplant from western Michigan's Euro-ethnic neighborhoods, Protestant churches, and ice hockey rinks, I knew little more about these earliest Americans than I did about the Indians of India. Now the assignment I had accepted would soon change all of that. Yet this rare opportunity just as easily might not have happened at all.

🦌 🦌 🦌 🦌 🦌

After a year in California, I was running on empty. Disenchantment magnified my yearning for the Rocky Mountains' familiar embrace. Unexpectedly one evening I received a phone call. It was a familiar and much-welcomed voice. As if throwing me a lifeline, the caller redirected my career.

"Hey, Bruce. How's California's cactus and lizards treatin' ya?" It was Bob Phillips, a wildlife scientist with the U.S. Fish and Wildlife Service (USFWS) and my previous supervisor in Sheridan, Wyoming. Before accepting my first permanent appointment in Riverside, California, I shared my misgivings with Bob about how it might turn out.

"Funny you should ask, Bob. It's been pretty frustrating lately."

Stifling the urge to unload, I caught up on the latest life changes of my former co-workers in Wyoming. Then anticipating my next questions, Bob filled me in on my former charges. In a few sentences, he detailed how many deer, coyotes, grouse, eagles, and a menagerie of other wild critters remained "beeping" on the air. To me they were like far-off children and I missed tracking their radio transmitters and learning their whereabouts and fate amidst a burgeoning maze of coal strip mines and haul roads.

"Deer seem to be adjusting, but sage-grouse, well not so much."

When he paused, my frustrations rolled out like water from a ruptured dam.

"After almost a year, I've come to a roadblock trying to write a management plan for the Santa Rosa Mountains—trying to protect their desert bighorn sheep. So what if I've produced a nifty habitat map of the Santa Rosas. It's just lots of pretty plant communities, with one big problem. It lacks integration of wildlife's biological needs. It just seems academic."

"I hear you," Bob said. "You know from your work up here that's the linchpin. Needing to know what's important to the animals and why."

"Yeah, and my supervisor's response when I explained the shortcomings to him was, 'Just write the plan with whatever you have.'"

"Sounds like getting a final product's more important than what's in it."

"That's right," I emphatically agreed. "I can no more write a useful plan without data than build a house without lumber." Hearing the insistence in my voice, I realized how much emotion I had bottled. "The Santa Rosas are an amazing place. The mountains and their bighorn sheep deserve better," I added as an epitaph to my feelings of defeat.

The line was quiet for a long moment. "I sympathize with you, Bruce," he started. "Wildlife needs more of your kind of passion."

In an instant, these words snapped me back to the present. "Gosh, I've been on a rant. Sorry, Bob. Seems like I haven't tried to talk this out with anyone else." Having said the words, I recalled that he had been the kind of supervisor that gave his employees plenty of talking space.

"So, I haven't even asked. Is there something special you called about? Something I can do for you?" I asked.

"Maybe it's what I can do for you." Rising at the end, his voice betrayed that he had called with a purpose, and not just to shoot the breeze.

"I'll get right to it. There's a job advertised in Lander. When I saw the announcement, I immediately thought of you," he continued. "In case you haven't seen it, I wanted to give you a heads-up."

"Lander," I repeated, trying to place the spot on a mental map of Wyoming. "What's it involve? Who's the job with?" I asked with instant interest. In two or three previous conversations since I had left, he avoided any talk about job vacancies, knowing how hard it was for me to leave Wyoming. This must be something special, I realized.

"It's a wildlife biologist position on the Wind River Indian Reservation. Mostly working with big game," he enthusiastically answered, then added the job was the first of its kind.

"And who's it with?" I eagerly responded.

"The Fish and Wildlife Service."

I read the self-assurance in his voice. He knew me well; well enough to know that I would respond like a shark smelling chum.

I first learned of coveted job opportunities in the USFWS from my graduate committee chair, Dr. Bart O'Gara. Bart was the assistant leader of the Montana Cooperative Wildlife Research Unit in Missoula, one of 37 such units administered by the USFWS. One of the more appealing aspects of the USFWS was its unique mission. It obliged scientists to work with wildlife populations *and* their habitats—an integrated approach to resource management. Nowhere is this more apparent than throughout the system of 550 (and counting) national wildlife refuges that the Service manages. This diverse collection across 50 states totals 96 million acres. It's essential to the nation's effort to sustain species that reside within our borders and others that migrate from the Arctic to Argentina.[1]

In other federal agencies, such as the U.S. Forest Service and Bureau of Land Management (BLM), biologists were largely restricted from handling animals, monitoring movements and survival, collecting samples for health analyses, etc. Such responsibilities were reserved by the individual states, and this was the hitch I had run into in California. Working with both the land and its animal populations was a holistic approach that intrigued me. I wanted to do big-picture work and make significant conservation contributions. Yes! I wanted to work for the USFWS.

At my urging, Bob highlighted the details for me. Considering my California experience, I couldn't help but wince at his first words, "*Develop a wildlife management plan* ..." But, I reminded myself, this was not the BLM. As he continued, "... population surveys of big game species across 2.2 million acres ... locate important winter ranges, reproduction areas, and migration routes," a surge of excitement swelled in my chest. This sounded good—almost too good to be true.

I thanked him again, we said our good-byes, and I hung up the phone at the dial tone. I sat in the silence of my living room with an onslaught of Technicolor thoughts bouncing from the walls. At first, of course, I had to reassure myself that this call had really happened. But quickly I felt a clarity about what it was that was drawing me there—the Wind River country—rather than keeping me here.

🦌 🦌 🦌 🦌 🦌

I have always had this enigmatic passion to tackle tasks I believed would make a lasting difference. In my fledgling career, I sought to channel that energy toward sustaining wildlife and the lands where they lived. I believe that's what left me feeling empty in California, with unfulfilled aspirations and fading hope.

In the 1970s, permanent wildlife jobs were as hard to land as a sturgeon on a fly rod. So when an offer came in May 1977, I quickly accepted even though working in the desert for the BLM wasn't on my radar screen. After I had worked an itinerant succession of seasonal jobs, the BLM offered me a chance to see a project through, and the position's focus appealed to me. I was charged with crafting a habitat management plan for the Santa Rosa mountain range and its featured species, the desert bighorn sheep. A mountain-dwelling large mammal, the bighorn was ecologically similar to the mountain goats I had studied in Montana for my master's degree.

The bighorns were sparsely distributed across this rugged environment that rose from near sea level to 5,000 feet. To write the plan required mapping the mountains' habitats and determining which ones the sheep used most throughout the year. It was longstanding dogma that sheep visited water holes habitually. But I found little data to evaluate if bands of bighorns used one, or several, of the scattered seeps and springs that sustained them—virtual fonts of life during brutal summers when temperatures nudged 120 degrees. If some were wide-ranging, then linkages between water holes were as important to protect as the water hole environs themselves. Isolation of subpopulations could lead to inbreeding and exclusion from seasonal resources. That could threaten the persistence of the overall population. I suggested to the BLM and state of California officials that these matters could be resolved with proper study.

After several weeks of tromping sand washes and rocky ridges to visit water holes and map vegetation, one thing became clear. Better technology than binoculars and boots was required. Buff-colored sheep were well camouflaged and thinly sprinkled across a tapestry of crags, sand, and surprisingly lush vegetation. Amassing unbiased observations of movements and behavior (in between my job's other pesky duties) was out of the question. To understand their travels required radio-monitoring some sheep.

Radio telemetry data would improve the BLM's management of the fifty-mile-long mountain range, and thereby benefit California's desert bighorns. Specifically, knowing in which habitats the sheep traveled and spent time would help prioritize private lands for acquisition—a pricey proposition in the backyards of Palm Springs, Palm Desert, and Rancho Mirage. Ownership of the Santa Rosa sheep range was a checkerboard of federal, state, railroad, and other privately held lands. Coordinated land protection among all landowners was paramount to conserving this largest remaining U.S. population of desert bighorns.

With my supervisor's dubious approval, I developed a study plan to capture and radio-collar a representative sample of sheep. I acknowledged the obvious. State biologists would conduct the captures. The BLM would coordinate the monitoring of radio collars. I quickly learned, however, how closely the state of California guarded their dominion over Santa Rosa bighorns. My proposal to radio 12 to 15 animals was promptly rebuffed. The reason, I was told—the BLM managed land, *not sheep*.

For my part, I was naïve. With the backing of the local state biologist, I thought the merit of my proposal would win over others. They would be "color-blind" to the agency that formulated it. I simply failed to understand that California viewed the BLM more skeptically than it did the other federal resource management agencies. That is why *other* agencies, not the BLM, managed our nation's premier public lands.

Originally, much of the western landscape was classified simply as "public domain" and administered by the General Land Office. As the western frontier was settled under the Homestead Act and Desert Land Entry Act of the late nineteenth century, the most productive and valuable lands were withdrawn for private ownership. From the remainder, national forests, parks, and wildlife refuges were carved out and placed under the appropriate administration. All remaining acres deemed least productive, least scenic, sparsely watered, relatively unforested, mostly rattlesnake infested—and presumably godforsaken—defaulted to the BLM after the agency was established in 1946. In other words, the BLM got the leftovers—what nobody else wanted. This was true also of most reservation lands where hundreds of remnant Indian tribes were confined during the late 1800s. Although some tribes were allotted reservations within their historic homelands, many were not. Jon Meacham writes that President Andrew Jackson believed "with all his heart" in the removal of all Indians from east of the Mississippi. "He was as ferocious in inflicting harm on *a* people as he often was in defending the rights of those he thought of as *the* people." To those who argued for Indian rights, Jackson responded that removal would "guarantee the survival of the tribes, which would otherwise be wiped out," asserting that "coexistence was impossible."[2]

Passage of the federal Indian Removal Act of 1830 began the uprooting of Indian tribes dwelling east of the Mississippi River for relocation to "Indian Country" to the west. Westward expansion of Euro-American society further displaced relocated tribes and resident nations from western homelands. Reservation treaties with Indian nations were modified, diminished, and abrogated by the federal government to accommodate white settlers and economic enterprise.

By default, the BLM became responsible for 258 million acres of land, or 13 percent of the United States. Ironically, beneath the semiarid and austere surface of those lands lie huge quantities of coal, oil, natural gas, coal bed methane, and hard rock minerals of untold value. In 2007, the BLM's mineral leasing activities returned about $4.5 billion in receipts from royalties, bonuses, and rentals to the U.S. Treasury. Beyond overseeing more energy and mineral extraction, the BLM administers more livestock grazing permits—over 18,000—than any other federal agency. Not surprisingly, the BLM has been labeled, among other disparaging nicknames, the "Bureau of Livestock and Mining." Others prefer the "Bureau of Latent Management."

Perennially understaffed and underfunded, the BLM struggled to do more than issue permits for resource extraction, use, and abuse. Its lax oversight and enforcement were roundly ridiculed by conservationists and other agencies alike. With the enactment of the Federal Land Policy and Management Act in 1976—a year before I hired on—greater statutory emphasis was placed on multiple use and land stewardship. This was supposed to elevate conservation efforts for wildlife, watersheds, and wilderness, through planning efforts like mine. However, funding shortfalls, an archaic agency mindset, and America's hunger for energy and "outdoor playgrounds" still undermined (pun intended) the agency's efforts to conserve wildlife and wildlands. Thus, the BLM struggled as the stepchild among federal resource agencies. As I also found out, "The BLM doesn't do wildlife research."

In the space of 20 years, California's desert bighorns crashed in the Santa Rosas and adjacent peninsular ranges to the south. Numbering between 900 and 1,000 in the 1970s when I was there, they plummeted to 300 animals by the mid-1990s. The survivors were subsequently listed as federally endangered in 1998.[3] Although all causes for the decline were unclear, land development, degradation, and fragmentation were paramount. As the population tanked during the 1980s and '90s, ecological studies of the sheep were launched. Some used radio telemetry and geographic information systems to identify bighorn movements and critical habitats—the very work I had proposed a decade before.

Perhaps these efforts and ensuing land protection came too late. Or was the sheep population fated to an inexorable march toward failure? Either way, proactive planning and conservation are always more effective than reactive mitigation. The 1998 species recovery plan for the peninsular desert bighorn carried a $73 million price tag. That was the population's estimated rescue cost so it could be removed from the threatened and endangered species list.[3]

I learned hard lessons from my California experience about the foibles of science as taught in college. Academic training and technical skills can carry conservation only so far. Biologists must be persuasive in communicating, even promoting, their ideas from resource needs to research results. I failed to garner enough support from both professional and citizen wildlife interests. Yet because there was no conservation crisis in the 1970s, it may have made no difference. By comparison, what so intrigued me about the Wind River possibility (besides escaping the traffic, crime, and smog of Riverside) was that tribal governments on Indian reservations control both aspects of the wildlife resource: habitat and animals.

🦌 🦌 🦌 🦌 🦌

The day after my conversation with Bob, I found the job announcement among the "green sheets" posted at the BLM office. I polished my job application that evening and mailed it the next morning. After two telephone interviews with Dick Baldes, the Lander field station's project leader, I received a third call. I had been selected to be the first wildlife biologist to work on Wyoming's Wind River Indian Reservation (WRIR). I was overjoyed! The USFWS wanted to hire me. It took only a breath to accept the offer.

Actually, I conditionally accepted. I wanted first to visit the Lander office to meet Dick but also to talk with tribal officials.

The field office in Lander was established in 1961 as a Fisheries Technical Assistance Station. Since 1974, Dick and an assistant biologist had managed the Shoshone and Arapaho tribes' wild fisheries. Dick was tan, in his late 30s, and built like a fireplug. As an American Indian raised on WRIR, he had dreamed of this assignment since his college years. A ready smile accompanying the enthusiastic voice I had heard on the phone greeted my arrival in May 1978. From the moment I met him he reminded me of golfer Lee Trevino, the "Merry Mex."

Dick showed me the document that made this opportunity possible after two decades of fisheries support. The USFWS was honoring a year-old appeal for further assistance by the reservation's government. Tribal Resolution Number 3923 requested "a wildlife biologist position in Lander to assist in collecting data to protect habitat and wildlife, and to manage and insure the optimum potential of wildlife species on the Wind River Indian Reservation now, and for the future."

Most of the first day we spent on the reservation. As we drove miles of pavement, gravel, and dirt exploring slices of the land, Dick chattered nonstop about his work and the conservation challenges the Indian tribes were facing: stream dewatering, contested water rights, economic woes, and, of course, stressed wildlife. From flanking mountain ranges to the north and west, the Wind River Basin spread like a great sea of sagebrush and grass. Here and there sandstone buttes poked from its serene olive surface like tawny oceanic atolls. This was a big place, the kind of breathing space a person could get lost in, waiting for the future or searching for the past.

Though I had not been here before, the feeling was familiar. I had known a similar wonderment at the unfolding of prairie to peaks a decade before. On a road trip to check out the University of Montana in 1968, a friend and I saw the West for the first time. There was so much to absorb along the way: mounted cowboys and their dogs moving cattle, skittering pronghorns and big-eared mule deer, soaring mountains and snaking streams, a stockyard sign advertising "Used Cows," and birds—extravagant magpies and crested Steller's jays that I had never seen. With no travel guide beyond our Rand McNally, we anticipated with wild imagination what lay ahead. Now I felt a tinge of that same novelty in Wind River country—a place I had never been, but anticipated getting to know.

As we skirted the Wind River's greening cottonwoods, Dick fielded more of my questions about the reservation people's views about wildlife. He sketched Shoshone and Arapaho history and each tribe's cultural distinctions. With my mind overloaded and running low on questions, I abstractly asked, "Anything else?"

"Of course the Shoshones are the better-looking ones," he grinned.

I knew he was enrolled and this signaled his affiliation. Increasingly comfortable with our banter, I turned to him and asked, "So are you Arapaho or Shoshone?"

In a likewise jocular response, he invited, "Which one do you think?"

"Well, Shoshone of course!" And we joined in a long laugh.

"Good answer," he followed. "You may just work out."

🦌 🦌 🦌 🦌 🦌

Like most anyone, I had some qualms about the unknown. Dick had told me by phone that WRIR lacked hunting regulations. Game herds were spare and decreasing. I had made the two-day drive to Lander to assess the challenges of the job up close. Most importantly, I kept pondering, would the expectations and sincerity of tribal leaders persuade me there was a genuine chance for success? I would be leaving one job where success appeared elusive; I didn't want to trade places for more of the same.

Our first stop on the second day of my two-day visit got squarely to the heart of the matter. We drove the 20 miles from Lander to a former military post, now the town site of Fort Washakie. Dick parked before a modern, native-stone complex situated incongruently among former army barracks and other dated buildings.

"This is where the tribal government offices are. The tribal council meets in here," he explained.

I recall the anxious anticipation as we walked toward the building's front entrance. A larger-than-life bronze statue of the Shoshone Chief Washakie guarded the bank of glass doors. In molded plastic chairs in a tight conference room, Dick and I met with three Shoshone councilmen for an hour.

I had no doubt as we exchanged handshakes and probing looks that these men came from a culture and background far different from mine. Wes Martel, a sturdy man with jet black hair, was, like me, about 30. He and Robert Harris, Sr., the oldest of the three at 65, showed the high-cheek-boned, chiseled features familiar from classic western movies and TV shows. Chairman Harris was a commanding presence. Over six feet tall, he had penetrating eyes, thick, combed-back graying hair, a baritone voice, and an eloquent wit that suggested the self-assurance that had served

Dr. Frank Enos receiving Small Business of the Year award in 1965 from Mrs. Nellie Scott, Wind River Reservation Economic Development Planning Committee

his people well for 40 years as a tribal councilman. Flanked by Mr. Martel and Dick at one end of the rectangular table, he chatted for several minutes with the others about matters unrelated to why we were here. The third man, Frank Enos, looked less the part. He wore a rosy complexion topped with a brown crewcut. No Indian beaded buckle surrounded his ample waist. To many, only his Shoshone family name might give his Indian heritage away. Waiting for the formal meeting to begin, he sincerely thanked me for traveling to the reservation. He asked about my past work as a biologist. I in turn learned he was Dr. Enos. After years of large-animal veterinary practice, he now ran a family cattle operation on the reservation. I knew in a few minutes of relaxed conversation that I would enjoy working with this man.

Dick opened the discussion. At the outset he made sure everyone knew, "Bruce asked to meet with council members. He wants to learn about the reservation's wildlife condition directly from you before he decides to take the biologist job."

Dick had a habit of being a bit abrupt. I hoped his lead-in and my sudden awkwardness didn't translate into me putting on trial their considerable efforts to establish the wildlife position. To my relief, I didn't detect any negative reactions in their faces, probably because they knew, far better than I, what a daunting task lay ahead. As they swapped testimonials, I learned that the reservation had no, I repeat, *no* written record of wildlife populations. No population estimates, no information on distributions, no authoritative estimate of trends. They left no doubt; I would be starting from scratch.

"Do you have a species list?" I ventured, then tried to answer the blank stares. "A list of the different animals on the reservation." No, not even a species list.

"Of course, there's deer, elk, speed goats ..."

"Antelope," Dick interjected with a grin.

"... some bighorns and bear, and maybe still a few moose," Wes Martel continued. As he and Dick added sage-grouse and a half-dozen others of special significance, Chairman Harris interrupted, "It's the big game that are in trouble. That's what we need to find out about first."

I found it astonishing. Here was an area of 2.2 million acres, the size of Yellowstone National Park, that was an informational void. In Yellowstone nearly 400 species of mammals and birds were recorded; there could be just as many here. It was like stepping back a century in time. I would have to start with the most basic data gathering to build a conservation plan. Despite the lack of hard data, all three men shared a genuine concern, even distress, about wildlife conditions. The problem was obvious. "All you have to do is go hunting for an elk or deer to find that out."

From a folder Chairman Harris drew a slick-covered bound report. He slid it across the table to me. "With a federal grant," he summarized, "the Joint Business Council contracted this. It's a review of our Fish and Game Department. That's why we need a wildlife biologist here—and because all animals are sacred and the relations of the Shoshone people." These piercing words said more in seconds than a job announcement could convey in ten pages.

The Fish and Game Department served only law enforcement functions, and with precious few laws to enforce at that. Only selling meat, wanton waste (abandoning killed game in the field), and hunting with a spotlight at night were prohibited. Based largely on interviews with tribal game wardens, councilmen, and other knowledgeable individuals, the report I was leafing through concluded that "all species of big game are in trouble and decreasing on the Reservation due to excessive hunting by enrolled tribal members. Present low populations cannot be maintained without controls. In fact, moose and antelope (pronghorns) are at such low levels that continued hunting may eliminate these species ..."

This review by the Agricultural Resources Corporation of America listed the steps needed to rescue the reservation's wildlife. Foremost, WRIR needed systematic professional inventories of resident game populations, as well as guidance consistent with tribal members' attitudes on how to replenish game populations.[4] And there was the rub, I would learn later—reconciling long-standing cultural attitudes with modern-day conservation principles.

Our conversation concluded with a laundry list of challenges, including the lukewarm enthusiasm of the Arapaho Council (conspicuous to me by their absence here), and complacency or outright opposition to change by many members of both tribes. Two things rang clear. The reservation's wildlife needed urgent help, and these three leaders were committed to that.

I was taken by the sincerity of these men. As we rose from our chairs and Dick and I walked toward the door, Dr. Enos approached me. "I don't want to see the wapiti vanish as the bison have from the Wind River country. If it's necessary to completely close the hunting season for one or more years to restore big game populations, I will support that."

His stature as a veterinarian and chairman of the Tribal Fish and Game Committee (six tribal councilmen providing oversight of the Fish and Game Department's law enforcement officers) promised a key ally in the effort to recover wildlife herds. I came away feeling like a new world had opened to me. This was a once in a lifetime—no, once in a generation—opportunity. I relished the chance to know better these people and the wildlife that was so central to their heritage. A month later I was on my way to Lander, Wyoming, to undertake the biggest challenge of my young career.

🦌 🦌 🦌 🦌 🦌

The 1,100-mile drive from Wyoming to California went by in a flash. I replayed my two-day visit trying to memory-bank each detail. At each fuel and food stop I scratched more ideas in a spiral notebook, plotting how to organize this job I had accepted.

Two things excited me most, if you don't count leaving California. I divined the major difference between this opportunity and my present job. The BLM needed me to write a habitat management plan. It was a policy requirement. On the WRIR, the people—at least the ones I had met—wanted me to do the job, to help restore the wildlife so important in their lives. What motivation that provided! The prospect of the fieldwork also appealed to my personal interests. After studying mountain goats for three winters to earn my master's degree, every job I held dealt largely with researching or managing hoofed mammals, or ungulates, as the group is called. Whether it was deer and pronghorns I watched navigating pockmarked Wyoming mine lands, mountain goats I surveyed in Washington's North Cascades National Park, or elk I spied on in the high country of Montana, big animals and I seemed drawn together. The WRIR may have small populations, I conceded, but recovery potential was there. The land held every post-Pleistocene species except the buffalo and the wolf. And through the windshield of Dick's pickup, I witnessed a bounty of empty habitat. Back home I would ponder the reservation maps rolled on the seat beside me. I would contemplate how to refine the crudely penciled polygons used by big game as winter range. As another hundred miles of Nevada sagebrush slipped behind me, I thought more about the big-animal days that lay ahead.

🦌 🦌 🦌 🦌 🦌

There are pros and cons of work with any species. On the con side of the ledger, large mammals—be they herbivorous hoofed beasts like deer, moose, and mountain goats, or carnivorous ones like bears, cats, and wolves—are highly mobile and often migratory. Their study demands sophisticated methods to monitor movements and survival. Population surveys from aircraft, biotelemetry from airplanes or satellites, and large-scale habitat analyses are generally required. Such studies are logistically complex and accordingly expensive.

On the positive side, these are wonderful animals to spend time among. They inhabit some of the continent's most awesome wildlands, the kinds of big, raw landscapes that absorb me. Being large means that individuals, or their spoor, are more readily observed compared to the "little" species. That makes them particularly conducive to behavioral studies. Large animals on average outlive smaller ones, so scientists can develop detailed life histories over successive years and generations.

Importantly, humans readily relate to large mammals, compared to less charismatic creatures such as insects, snakes, or bats. We're a large mammal ourselves; our evolutionary journey was closely linked to and shaped by others. Some provided our species with food, clothing, tools, and other resources; some served as objects of fear and peril; some were significant competitors; still others became important symbols in our culture and spiritual beliefs.

Stephen Kellert, professor at the Yale School of Forestry, identified a typology of nine values humans associate with large mammals. They range from aesthetic to utilitarian, and symbolic to dominionistic.[5] In total, these values influence our attitudes and behaviors. They shape modern conservation of large mammals and their habitat. What value we place on a particular species or population reflects our experience, education, and culture. Differing values foster conflicting opinions about appropriate management. Nonetheless, large mammals seldom suffer from the apathy or anonymity relegated to species lacking fur, big flesh, and loveable faces. This just reflects the appeal of creatures bearing practical worth or resemblance to us. As George Orwell observed in *Animal Farm*, "All animals are equal; some are more equal than others."[6]

A recent analysis found that less than 21 percent of the earth's terrestrial surface still supports all of the large mammals—those weighing at least 44 pounds—that were here in 1500 AD.[7] Many species face growing threats. Fortunately, better knowledge can improve our stewardship of "umbrella" species such as bears, elk, and mountain goats that require vast or relatively undisturbed (by us) ranges to survive. Conserving populations of such charismatic species protects entire communities of biota that share their habitats and ranges. Reserving tracts of tropical Amazonian rainforest for jaguars, for example, protects thousands of far less well known species, and many others yet undiscovered. This concept was exciting to me. By working with large mammals, I could benefit whole ecosystems. As importantly, I would be working toward a common cause with the people I had met and many others I would come to know on the reservation.

2

On the Reservation

Wind River Indian Reservation is one of those places most people have only visited at 65 miles per hour. In 1978, it was home to 5,700 Shoshone and Arapaho Indians in a scattering of rural settlements and ranchlands. It remains sparsely populated today. The reservation is better known for where its roads lead than for its own attractions. From the east, it provides a gateway to Jackson Hole, Grand Teton National Park, and the south entrance of Yellowstone National Park. Lander, a welcoming town of 10,000 near the reservation's southern boundary, is home to the National Outdoor Leadership School, where thousands of students have learned mountaineering and survival skills. South beyond Lander stretches the Red Desert, rich in pioneer history and traversed by the Oregon Trail. To the east lie the towns of Thermopolis and Riverton. Thermopolis is home to the world's largest mineral hot springs, once a place of healing and restoration for Plains Indian tribes. Built on lands ceded from the reservation in 1906, Riverton's 9,000 inhabitants make it the largest town within the boundaries of WRIR by far.

Other reservation towns named Arapahoe, Ethete, and Crowheart could have both city limit signs on the same post. Yet WRIR is less desolate than some parts of Wyoming where drivers will change a flat tire doing 40 miles per hour just to get somewhere else. One old ranch hand characterized the Wyoming landscape this way: "It's all a bunch of nothing—wind and rattlesnakes—and so much of it you can't tell where you're going or where you've been, and it don't make much difference."[1]

Fort Washakie is the tribal headquarters. It draws its name from the Shoshone Chief Washakie. He successfully worked with the cavalry and negotiated the present reservation as a homeland for his people after it became clear that the end of their traditional lifestyle was at hand. The Shoshones were one of several tribes inhabiting the Great Basin and Columbia River Plateau region. The Little Ice Age, which began about 1350 AD, brought lingering drought to the Great Basin and increasing precipitation to the Great Plains. For the next two centuries, wildlife herds flourished on the Plains and declined farther west. Beginning about 1500 AD, significant numbers of Shoshones shadowed the game to the western reaches of the Plains.

These Shoshones obtained horses during the late seventeenth century.[2, 3] They and their close relations, the Comanche, were among the earliest-mounted nomadic buffalo hunters on the Plains. They held a clear advantage over other Plains tribes during the early eighteenth century and expanded their hunting territory eastward in pursuit of migratory herds.

By the late 1700s, all Plains tribes had horses. The Northern Arapahos obtained the horse about 1735 and migrated from northeastern Minnesota to command a central region of the Great Plains. Along with their allies the Cheyenne and Dakota, they often clashed with the Eastern Shoshones.[4] Like the Shoshones, the Arapahos lived by the buffalo economy. Buffalo meat became their diet staple. They fashioned bones into tools, and tanned and stitched hides into clothing, tepee covers, and robes for their lodges. The hides were also valuable trade items to obtain guns,

Shoshone Chief Washakie

knives, ammunition, copper kettles, etc. Swaddled at birth in the soft skin of the buffalo, these people would ever be touched by it, even in the winding-sheets of death. During the past millennium, agricultural crops—including maize, beans, squash, gourds, pumpkins, and sunflowers—were staples of American Indians living east of the Mississippi and in the American southwest.[5] Thus, a hallmark of the nomadic Plains tribes was their traditional hunter-gatherer lifestyle and greater dependence on big game for food.

As other Plains tribes acquired horses, the Shoshones lost their competitive edge. By 1740, the formidable Blackfeet to the north had acquired guns from French and English traders in Canada. Blackfeet and their allies had driven the Shoshones from extensive areas of the western Great Plains of southern Saskatchewan, Montana, and Wyoming by 1780.[2, 5] The Shoshones had withdrawn to the western slopes of the Rockies from Montana to Utah by the time Lewis and Clark made contact in 1805. White explorers and settlers variously called this core group Eastern Shoshones, Plains Shoshones, Washakie Shoshones, and Wind River Shoshones. The Shoshones called themselves Kutsindüka, the Buffalo Eaters.[2]

On July 2, 1863, Chief Washakie signed the Fort Bridger Treaty. This contract with the U.S. government set aside 44 million acres of Wyoming, Colorado, Idaho, Montana, and Utah—an area larger than Massachusetts—in which the Eastern Shoshones had lived and hunted for generations. By that time, the buffalo and other big game were under siege by the white man. For example, in his report to the Thirty-sixth Congress, Second Session of 1860, C.H. Miller, one of the engineers with F.W. Lander—who supervised construction of the Central Division of the Pacific Wagon Road (known as Lander's Cutoff), and after whom the town is named—stressed how this new immigrant road would bring added misery to the Indians:

The new road in many instances follows the summer and fall trail of the Shohonee tribe. The animals of the immigrants will destroy the grass in the valleys where the Indians have kept the pine timber and willows burnt out for years as halting places in going and coming from their great annual buffalo hunts, and I believe, even beyond the mere question of policy, that it would be a very unjust and cruel course of action for the government to pursue should we take the use of the land without reimbursement to the tribe.[6]

When Lander broached the subject of the new road to Chief Washakie, the Shoshone leader responded that:

Before the immigrants passed through his country, buffalo, elk and antelope could be seen upon all the hills. Now, when he looked for game, he saw only wagons with white tops and men riding upon their horses. He admitted that his people were very poor, that they had fallen back into the valleys of the mountains to dig roots and hunt what meat they could find for their little ones.[6]

Still, Washakie steadfastly vowed not to fight the white intruders and their ruin of the Shoshones' way of life. Instead, he sought to negotiate the best accord he could for his people. Although they did not receive all they were promised, the second Fort Bridger Treaty of 1868 settled the Shoshones in the wildlife-rich Wind River country of what is now central Wyoming. Three times more the U.S. government would diminish the size of the reservation, ultimately shrinking the Shoshone homeland to 2,268,000 acres.

As a final injustice, the Shoshones were forced to share the WRIR with their historic enemies, the Northern Arapahos. The Arapahos had been less cooperative with the U.S. government and received no reservation of their own. At the prodding of government officials, they approached their former enemy for reservation privileges. Washakie granted them temporary use of WRIR in 1878 until they could be relocated—something the government later refused. In 1938, Washington awarded the Shoshones a monetary judgment for land to accommodate the Arapaho Tribe. Both tribes have since "co-owned" the reservation.

By 1884, the buffalo were gone. Millions were extirpated from the continent by market hunters for tongues and robes, by homesteaders as they settled the West, by "sportsmen" traveling the transcontinental railways, by the military to provide food but also to deprive the Plains Indians of their lifeblood, and by disease contracted from livestock. In 1887, William Hornaday surveyed the remaining bison in the United States. His report to the Smithsonian Institution could account for only 541 head, all in private herds save 100 in Yellowstone National Park.

At the same time, the Plains tribes were hounded by the military, infected with smallpox and other diseases to which they lacked immunity, and coerced without recourse to surrender more and more of their land—and their dignity. In a final *coup de grâce*, the U.S. government forced proud nations into a sedentary lifestyle utterly foreign to these nomadic hunter-gatherers. The General Allotment Act of 1887, championed by Senator Henry Dawes of Massachusetts, sanctioned the division of communal reservation lands into quarter sections (160 acres). Parcels were assigned to the heads of households. What remained afterward was opened to white settlement. Extensive acreage subsequently transferred to non-Indians, further diminishing most reservations.

The concept of *owning* land was as alien to the Shoshones and Arapahos as being forced to become farmers, to adopt Anglo names, and to abandon their native languages in favor of English. The Wind River country had been a wildlife stronghold, but the buffalo and elk had always been migratory. Unable to follow the game beyond reservation borders without reprisals by the government, the tribe lost its way of life. Asked about farming on the WRIR, an aging Chief Washakie snapped, "God damn a potato!"[7]

During the early reservation decades, Shoshones and Arapahos suffered starvation, high infant mortality, and social upheaval. Tribal members still trusted native ways to keep body and soul united, not white "civilization." By 1900, the year of Washakie's death and seven years after the death of Arapaho leader Chief Black Coal, WRIR evoked memories of pre-reservation years. Historic enemies shared a common fate and connection to the past. Yet, as Henry Stamm detailed in his book *People of the Wind River*, they competed for political, economic, and spiritual resources from the land and waters of the Wind River valley.[2]

🦌 🦌 🦌 🦌 🦌

I felt at once privileged, eager, and intimidated by the daunting assignment I had accepted: reconciling tribal traditions of unregulated, year-round hunting with recovery of dwindling wildlife numbers. The job promised to be one of restoring wild nature, but also one of social engineering. Noted author of *Nature's Restoration*, Peter Friederici, frames the mission this way: "Restoration is the best way to maintain the integrity of the natural world, and it is becoming a powerful social movement, precisely because it offers a chance to use human energies in ways as deeply satisfying as building a dam or cathedral. It perfectly links individual action with the needs of something greater, both in human society and in nature."[8]

In the annals of restoration ecology, this wasn't a monumental endeavor. The task didn't match returning natural streamflows and riparian function to the Colorado River, or recovering the American chestnut throughout Appalachia, or restoring hydrologic function, water quality, and native biodiversity to Florida's Everglades. Still, it presented a unique challenge—bridging the divide between modern conservation principles and native traditions and beliefs. Maybe the biggest obstacle to recovering wildlife on WRIR would be overcoming a century of distrust of U.S. government policy toward native peoples.

I hadn't worked with American Indians before. The only ones I had known were two snipers assigned to my battalion in Vietnam. I don't recall their names or tribe anymore. I do recall both were impassive and difficult to engage in conversation. They were also exceptional Marines. Between them, they had over 100 confirmed enemy kills.

Hunting by tribal members on the WRIR remained much as it had for centuries, with important exceptions. Modern rifles, 4 x 4 vehicles, snowmobiles, and other trappings of modern society gave hunters a decided advantage over traditional weapons and horse travel. By treaty, hunting and fishing were the vested rights of the enrolled members of the two tribes. If you were enrolled—requiring verification of one-fourth Shoshone or Arapaho blood—and 14 years old, you had a vote. Any diminishment or modification of vested rights required a majority vote of both General Councils, comprising all enrolled tribal members—*both* being the key word.

I found that grievances persisted between the tribes. Shoshones resented having to share their land and resources. Arapahos resented that as the more populous tribe, they had only an equal say on issues of vested rights. Issues of common jurisdiction were addressed by the Joint Business Council, which combined the Arapaho and Shoshone tribal councils of six elected members each. At some meetings the tension was palpable. Suspicions of motive sometimes led one tribe to oppose the actions of the other—something our government's political parties are also fond of doing.

From time to time I attended these meetings. Sometimes I was there to present information. Particularly in the first year or two, the Council members were eager to hear the results of our recent big game surveys. Other times I alerted them to environmental issues that required their attention, such as the surface water contamination I had discovered that was caused by oil well effluents. Although I looked forward to these collaborative encounters, it took time to nurture common trust and reliance. I recall my early unease, particularly at that initial meeting.

Dick scheduled a time slot in the Council's first weekly meeting after my arrival on the job in June 1978. I had met with three Shoshone councilmen at the sprawling stone complex at Fort Washakie less than a month before. Now I would meet the entire Joint Business Council. Arriving early, I scanned the glass cases that graced the far wall of the foyer and housed historical highlights of the fort and of the Shoshone and Arapaho people. Corridors led right and left from the foyer to various business offices, including the Tribal Fish and Game office. Between the glass display cases, a pair of oak doors led to the inner sanctum—the Council chambers. Beyond those, I could only imagine what awaited me in my first formal meeting with the Arapaho and Shoshone leaders.

Dick accompanied me, as he did at many meetings with tribal officials. An enrolled Shoshone himself, he gave our office and me—because he hired me—an intrinsic legitimacy. Psychologically, it eased my entry into tribal politics—and psychology is everything. As Lee Trevino once quipped about mastering the game of golf, "It's 90 percent mental, and the rest is in your head."

We entered the chambers at our scheduled time. Dick chose seats to the right of a central aisle, halfway forward among several rows of plastic-seated metal chairs. The room was voluminous, accentuated by a stage in front. In russet leather and bar-tacked chairs behind an arc of heavy-legged desks sat a dozen stern-faced men, all Indians of course. The head of a bighorn ram and another of a huge mule deer sporting seven points on one side and nine on the other ruled the wall behind the councilmen. Four floor-to-ceiling windows and recessed lighting in the fifteen-foot ceiling illuminated the chamber. The room seemed still larger with only a handful of others conducting business that day. It reminded me of a courtroom where I had appeared to protest a traffic citation in California. I recalled that had not ended well.

We listened as the Council debated a business item. Three members fired questions at a man seated in the audience. I surmised he was an official with the Bureau of Indian Affairs (BIA). I don't recall the subject of the discussion, but I remember the tone. It was serious, almost tense. The grilling continued for several more minutes. I felt my stomach tighten, though as far as I knew I was present only to formally say "howdy."

My unease lifted shortly. After dispatching the BIA official, Bob Harris, the square-jawed Joint Business Council chairman whose penetrating eyes and imposing presence I had first felt last month, focused on us. "What have you for us today, Chicano?"

Without irritation, Dick grinned and replied, "I want to introduce our new wildlife biologist to the Council members."

Dick was part Mexican as well as Shoshone. The Chairman's remark, I would learn, was his way of keeping people in their place, of testing their forbearance. Outside the Council chambers, a rejoinder may have followed, but not in here.

Such repartee was not limited to Bob Harris. Starr Weed, a Shoshone councilman with mischievous eyes, was particularly fond of needling folks. At a subsequent Joint Business Council meeting, Starr wryly posed a rhetorical question to Dick that ended with the epithet "minnow." Whether this was simply an acknowledgment of Dick's fishery expertise or his vertically challenged physique, I'm not sure. But Starr could receive as well as dish out. I learned that such teasing was not from the well of ridicule, but from the fountain of affection. It was good-natured and a measured comeback was taken as such. I came to welcome this banter with councilmen, as it was a sign of acceptance. Curiously, I don't recall any tag applied to me; certainly nothing comparable to "minnow." I suppose this average-height, all-white wildlifer could never achieve Dick's level of approval.

At this first of many meetings to come, I rose as Dick introduced me to the Council. Twelve pair of dark eyes converged on me as I waited for any comments or questions they might have. Instead, hollow silence crystallized the void between their world and mine. After several uncomfortably long seconds, I announced, "I'm pleased to be here at Wind River Reservation. I'm looking forward to working with each of you."

"Anything else?" Chairman Harris directed at Dick.

"No, that's all today," he replied and we retreated through the chamber doors.

This would take some time, I now clearly understood. I would have to prove myself more than in previous environments. I would rely on the products of my fieldwork and the key alliances I would develop to break the ice and cultivate constructive relationships.

🦌 🦌 🦌 🦌 🦌

During those first days on the job, I juggled ordering equipment (binoculars, spotting scope, and field supplies) with meeting contacts in state and local government. I also began learning the lay of the reservation's geographic and political landscape.

The Wind River bisects the reservation diagonally from northwest to southeast. Then it loops straight north again

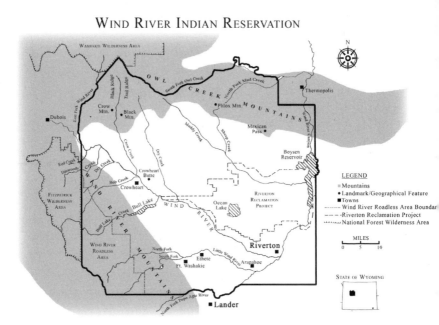

along the eastern border where its waters are impounded behind Boysen Dam. After Boysen Reservoir's water plunges northward twelve miles through Wind River Canyon, it exits the reservation no longer as the Wind, but as the Bighorn

River, like a street changing names upon leaving the city limits. From 4,500 feet elevation in the Wind River Basin, the land swells above 12,000 feet along the spine of the Continental Divide. Here the ramparts of the Wind River Range guard the reservation's western boundary. The Owl Creek Mountains, less rugged and born of sedimentary deposits rather than the Wind River Range's granitic bones, trends west from Wind River Canyon and joins the Absaroka Range at the reservation's northwest corner.

These two ranges provided refuge for most of WRIR's remaining game, save the remnant pronghorn population, which was scattered across the sagebrush desert and rolling foothills that cover half the reservation.

Pronghorns evolved in North America, whereas moose, elk, bighorn sheep, and most other extant large mammals are immigrants from Eurasia during the Pleistocene Ice Age. When Lewis and Clark encountered these small prairie speedsters, they were reminded of Old World gazelles. Accordingly, they called them antelope. Because this North American novelty is the only surviving member of its taxonomic family, *pronghorn*—a descriptive reference to its uniquely forked horns—is a more accurate name. Variously called "prairie goats," "speed goats," and "prairie ghosts" by western settlers, pronghorns reach speeds of 60 miles per hour, second only to the cheetah among land mammals. Their bulging eyes and fleet-footedness evince a coevolved history with a lightning-fast predator. Pronghorns can thank the New World cheetah's extinction some 13,000 years ago for a twenty-mile-per-hour advantage over its next swiftest predator, the wolf. For hunters like me who relish its delicate meat, the pronghorn endures as Wyoming's homegrown "fast food."

🦌 🦌 🦌 🦌 🦌

In my first weeks, I developed a two-part plan to reconstruct the history of each big game species on the reservation. First, I sought out elders of both tribes, those most knowledgeable about past numbers and distributions of each species. I interviewed those who were willing at their homes. Since there was no written record, their memories and information passed down through family members were my library. Additionally, I considered mailing questionnaires to a sample of tribal members. A concise list of questions would query their knowledge on wildlife matters. *They would be eager to tell all they knew, right?*

Next, I developed estimates—predictions if you will—of the WRIR's potential to support each big game species. I relied on two lines of information to estimate these "carrying capacities" of reservation habitats. In the northern Rockies, winter weather squeezes animals through a population-limiting bottleneck, at least those that don't take winged vacations to the southwestern U.S. or beyond. First, I researched densities of animals on similar winter habitats near the reservation. For that, wildlife biologists and wardens working for Wyoming's Game and Fish Department were my primary sources.

Second, I planned to measure how much vegetation grew and was eaten on winter ranges of the Owl Creek Mountains and Wind River Range. I geared this effort toward elk and deer forage. Because the staple bison were now gone, elk and deer had replaced them as the species of greatest importance to Shoshones and Arapahos. My earliest surveys indicated mule deer largely overlapped the winter distribution of elk. Just a scattering of mulies and white-tailed deer skulked the fertile bottomlands along the Wind River and its tributaries. Most remained out of sight in daylight or kept to lands deeded to non-Indians where tribal members were generally unwelcome. The scarcity of whitetails struck me especially. Content in our farm fields and woodlots, backyards and gardens, an estimated 26 million roamed the continent by the late twentieth century.[9] Like coyotes and crows, they are perfectly compatible, even commensal, with contemporary society. Though current species numbers rival those of precolonial days, they were nearly as rare on WRIR as Chinese restaurants.

Of the mountain-dwelling ungulates, elk were the easiest to observe and therefore to map their winter distribution. My initial aerial surveys of elk confirmed, alarmingly, what some had told me. Elk wholly avoided the lower slopes of the mountains that encompassed vast acreages of potential winter range. These foothills of grass and sagebrush provided windswept foraging sites and sheltered bed sites in thickets of juniper. Compared to the rugged and snow-laden higher ridges, foothills were more accessible to 4 x 4 and snowmobile traffic. They were winter range, all right, but only if elk and deer felt secure enough to venture there. I sensed tremendous opportunity to recover herd numbers, given sensible protection from hunting and harassment.

The remaining hoofed game, moose and bighorn sheep, were exceedingly scarce. Rarely did moose roam stream corridors beyond their mountain refuge. Such a mistake ended badly for the moose—a one way trip to someone's Frigidaire.

Remnant bighorn populations hung on in both mountain ranges. Even in winter, the sheep clung to the highest ridges and peaks in the windswept Owl Creeks. Lower cliffs and crests in the Wind Rivers, spared the deep snows where sheep grazed in summer, exposed bighorns to hunters during winter and spring. This held their numbers far below the thousands that ancestral Shoshoneans, the "Sheep Eaters," once hunted from rock hides. I twice chanced upon these relic ambush sites still clinging to ridges of ginger sandstone. Like Anasazi granaries, they memorialized a past culture. Standing where these people had stood, I tried to see the land—their world—as their eyes had perceived it centuries before.

🦌 🦌 🦌 🦌 🦌

Estimating the number of animals the reservation could support was a huge undertaking. It required extensive population surveys from aircraft several times each winter, mapping and categorizing habitats across winter ranges, and conducting labor-intensive inventories of forage production and use. For the browsers, I canvassed stringers of shrublands, measuring plant size and abundance along ridgelines and stream courses. Largely grazers, the elk's food supply was easier to measure than that of browsing deer and moose.

I was blessed with dedicated seasonal employees Kevin Berner and Tad Day, and American Indian field assistants Ray Nation and Philip Mesteth. Together, we collected reams of vegetation data and cranked out numerical estimates of food availability and its use, and ultimately of potential elk and deer populations.

At first glance, the task of estimating how much forage grew on mountain winter ranges seemed overwhelming. Where and how should I measure the vegetation? I began by partitioning the job into geographic units. The sampling protocols I adopted from textbooks and federal agency manuals. As if crafting instructions for Christmas toy assembly, I developed this stepwise approach to our work in the Wind River Range in 1979 and the Owl Creek Mountains in 1981:

1. I outlined broad areas to sample based on where I saw elk during winter surveys.
2. Throughout this mapped area, I randomly selected sections (square-mile units).

3. I stratified each section by habitat using features like terrain aspect, soil type, and dominant vegetation that I gleaned from topographic maps, soil maps, and aerial photographs in the WRIR's Bureau of Indian Affairs office.
4. Within the 30 habitats identified, I randomly selected map coordinates as locations to sample vegetation.
5. At each of these locations, I marked both ends of a 100-meter-long transect with a steel pin.

Although this framework's detail seemed extreme, even to my assistants and me, I wanted results that I could trust. As in carefully designed public opinion polls, painstaking stratification, randomization, and unbiased sampling were necessary to minimize our margin of error. With this blueprint completed, we headed to the field.

Sampling the grasses, sedges, and forbs (herbaceous broadleaf plants) eaten by elk was low-tech labor. Armed with grass shears, paper lunch bags, small spring scales, and clipboards stuffed with data sheets, we hiked to the starting coordinates of each transect. At ten locations equally spaced along each 100-meter transect, we established a circular plot. A length of surgical tubing, easily joined or disconnected by inserting a pipe fitting into the ends, delimited each plot. Simulating a grazing elk, we clipped each type of plant to ground level, bagged the material, and weighed the contents in grams—a finer scale of measurement than the English system affords. This was repeated for each plant species on each plot. The chosen plot area of 1.92 square feet facilitated mathematical conversions. Multiplying by 50 converted grams of vegetation per plot to pounds per acre of bluebunch wheatgrass, Indian rice grass, silky lupine, locoweed, etc. A season's fieldwork yielded impressive

Biological technicians Philip Mesteth and Ray Nation clipping vegetation, May 1980

columns of figures in a staggering number of tables. Yet this was far from the final answer.

At the office, number crunching began in earnest to estimate total food supply. The process went something like this: for all locations sampled in a given habitat (sagebrush for example), we averaged all pounds/acre measurements, multiplied by total acres in that habitat, repeated these steps for each habitat, and summed for all habitats. *Voilà!* Total food available to elk from November through April. Each elk required about 10 pounds of food per day. At this daily "meal rate" I could crudely estimate how many elk the food supply could support. The needs of other grazers were likewise taken into account. To avoid overgrazing the range, I made a final adjustment by reducing elk numbers by 50 percent. This left half the forage to feed lesser herbivores or decompose to soil humus, rather than be completely mowed and bagged as if by a Toro.

To corroborate this elk population estimate, I measured how much of the food supply was being consumed. Consumption was the difference between forage on plots protected by dozens of small wire cages (that excluded grazing) and forage remaining on adjacent unprotected plots. If 25 percent of the forage was eaten, the food supply could sustain twice the current number of animals (assuming my allowable offtake of 50 percent of forage). This assessment accounted for food eaten by all herbivores, including insects, small mammals, and livestock. Possibly 800 to 1,000 wild horses roamed nearly 400 square miles of elk winter range (240 square miles in the Wind Rivers and 150 in the Owl Creeks). No cattle grazed these ranges in winter, but in summer 22,000 head cropped grasses across reservation lands.[10]

Cattle roamed habitats least used by deer and elk. This I determined much as I had estimated elk distributions in Montana's Lolo National Forest in my college days—pellet group counts. On the plots where we clipped vegetation, my colleagues and I counted cow pies, heaps of horse apples, and groups of fecal pellets dropped by deer, elk, and moose. I used these measurements to learn which animals preferred which habitats and where competition for food was greatest. Cattle, for instance, spent most of their time in riparian habitats—among the habitats least used by mule deer and elk. Elk and deer had their favorites, too. Elk preferred sagebrush and grasslands, while deer preferred shrublands of bitterbrush (a low bush that browsers find tasty, despite the name) and conifer forests. These behavioral patterns distributed foraging across the land and reduced competition between species—evolution's exacting plan: no two species sharing the same niche in the same place.

But true habitat preferences of the wild ungulates these were not. Preference presumes choosing to use certain habitats among all those available. In the case of WRIR, wildlife choices were skewed by an overriding need for security, to remain out of sight—or at least as far from humans as feasible—to avoid being killed twelve months a year. Thus, miles and miles of mountain foothills remained all but devoid of game, even when snow buried forage on the higher mountain slopes. Our data simply indicated which habitats deer, elk, and moose selected from a diminished menu of choices.

🦌 🦌 🦌 🦌 🦌

If you're bored to tears with the details of this work, imagine the staying power required to see it through. Let me make clear—*vegetation sampling is truly monotonous work*. Clip, bag, weigh, record; clip, bag, weigh, record; a bag for each species at each plot on each transect in each habitat. Such brainless tedium is evidence that science is afoot. Fortunately, after I established the sampling design and protocols, my field assistants did most of the clip, clip, clipping.

I have always considered shears, or scissors, a tool designed for someone else. They just don't fit my hands right. Hammers, saws, and power tools are more my speed. Clipping vegetation reminded me of my sisters, Norma and Sandy, three and six years younger than me, sitting on their bedroom floor cutting out paper-doll clothes. If you're old enough, you'll remember that the dolls were cardboard. The clothes were paper designs on the pages of 8½ x 11 inch booklets—dresses, skirts, blouses, swimwear—complete wardrobes in a book. Norma would show her younger sister how to manipulate the silver shears around a ball gown's contours. She tutored Sandy to take special care to clip around the rectangular tabs that would fold each garment onto the doll, securing it in place. I could return from two hours of checking redwing blackbird nests in the four-acre cattail marsh behind our house, my German shepherd in tow, and they would still be clipping away.

"Want to help?" Norma would giggle as I peered into their room.

"No thanks," I'd reply, not wanting to deprive them of liberating a last pair of slacks from the booklet.

I suppose they considered my curious preoccupation with mapping bird nests, counting eggs, and recording hatchlings as humdrum as I did their paper-doll cutouts. They were mystified why I would slog through marsh muck to find muskrat houses—mounds of decaying cattail and bulrush with oversized rats nestled within. To me it wasn't boring, smelly, or pointless. It was fun and absorbing as the hours flew by. Little did I know that ecologist Paul Errington was simultaneously plying other Midwest wetlands, recording for posterity the "lessons as well as the beauties of marshes awaiting the perceptive" in *Of Men and Marshes*.[11]

My meticulous record keeping wasn't drudgery either—anything but. It was rewarding, like keeping box scores, batting averages, and ERAs of my team, the one that competed against teams of other neighborhood kids on my homemade baseball board game. These were ledgers to pore over and compare from year to year—wellsprings of mysterious patterns and trends.

My rounds with redwings were my first unpretentious dabbling in science. Not the noblest of subjects and deemed marauding pests by many farmers, the raucous birds were obligingly abundant and close at hand. Just minutes from our house awaited the marsh that rimmed Muskegon Lake. With spring's arrival, new nests interlaced clumps of cattail stalks. As my hip boots again raised the organic marsh scent, I wondered: were these the same pairs that wove last year's nests? How long did the birds live? To what state or country did they fly to escape winter? And were last year's hatchlings now busily incubating their own eggs here?

Although I fished and played sports and makeshift games with other neighborhood boys, the marsh was often my own. No one else inquired about the birds. I alone was the keeper of their annual account. As shown by the *Autobiography of Charles Darwin*, E.O. Wilson's *Naturalist*, and the unwritten memoirs of hundreds more, a quiet marsh or deserted woodlot has proven fertile preparatory ground for many a youngster's science career.[12, 13]

As I reflect on it now, my boyhood fascination with observation and note-taking prepared and perhaps predisposed me to a career in science. I realize also, it was not all my own doing. We all model behaviors of others, perhaps most in our formative years. Could it be that my mother's bookkeeping talents sifted into my nature? Or that my father's meticulous care and labeling of his roses (with me as the reluctant apprentice) founded my own attention to detail? Only now with both long gone from this world do I accord them this credit. I failed to see it, and much else, beneath the weight of the times. But time wilts pain and obscures scars—mine from a turbulent childhood where alcoholism's capriciousness ruled the household, regimentation became a coping skill, and woodland and marsh offered escape and harmony.

Mindlessly, I endured the tedium of the grass shears. As necessary means to an end, they were a tool of the habitat biologist as much as binoculars were a tool of the population biologist. Both produced data.

Like cataloguing blackbirds and muskrats, sitting on dirt shearing vegetation had a tangential appeal. It happened outdoors. Each field day provided fresh experiences beyond hiking to sampling locations and performing our Edward Scissorhands routine: a new redtail hawk nest; a badger flushing ground squirrels from underground chambers; a merlin streaking past in pursuit of a mountain bluebird—the chase culminating in a cerulean shower of feathers. At one site high in the Wind River Range, fossilized clams littered the ground. *We were sampling the floor of a great inland sea!* Where bivalves once filtered plankton from seawater, coyotes and Brewer's sparrows now foraged among sagebrush. What mysteries lay entombed within the layers of limestone beneath, we could only imagine.

Both Kevin and Tad became accomplished field botanists, but Tad was *really* into plants. I can recall him bursting into our office in Lander late on Friday afternoons. After four nights of backcountry camping, usually at dry campsites, the aroma foretold his arrival.

"Guess what new plants we identified this week?" Tad would exclaim.

Almost before I could answer, he would produce the delicate specimens he had pressed.

Among the field guides to plant taxonomy that Tad toted everywhere was my *Flora of the Pacific Northwest*. The 7 x 10 inch hard cover was barely bound to its 730 pages by the end of the field season in September 1981. His species lists for winter ranges in the Owl Creek Mountains included 22 trees and shrubs, 43 species of grasses and grasslike plants, and 139 species of forbs—all from elevations above 7,800 feet.

Working at high elevations presented some special challenges. During one outing in the Owl Creek Mountains, Kevin and Philip awoke before dawn to find their dome tent draped in snow. It fell as wet slush and froze to the fabric. The tent's frame strained and twisted beneath the weight, then collapsed with them inside. Extracting themselves at first light, they determined that sampling vegetation was problematic and returned to Lander. Kevin recalls that I intimated they were "insufficiently motivated."

"You could have stayed in the field knowing the snow would soon melt," he recalls me saying. Now that seems harsh and totally out of character for me.

"It seemed to me a perfectly reasonable expectation in mid-summer," I jokingly told him during a recent phone call. "In retrospect, I should have supplied you with shovels. You know, to avoid the delay of waiting for the snow to melt." Tad went on to earn master's and Ph.D. degrees in plant ecology. A professor in the Department of Botany at Arizona State University since 1995, he studies the influence of climate change on plants of the Antarctic tundra and Sonoran Desert. Kevin researched competition between white-tailed and mule deer in Montana's Swan Valley. Following his graduate education, he returned to wildlife work in Wyoming, then accepted a faculty appointment at the State University of New York at Cobleskill where he teaches wildlife and ecology courses. Philip, a Sioux Indian, is now the facilities manager at the Wind River BIA Agency. Ray earned a B.S. degree at the University of Wyoming. After three years at our USFWS office, he served in several positions at the Wind River Agency. He is now the assistant superintendent.

Like Tad making a botanical list for the Owl Creek Mountains, I kept lists from my wildlife inventories. In four years, I recorded 31 mammals and 162 bird species on WRIR. Another 35 mammals and 80 bird species potentially used this grand landscape, either year-round or seasonally. This original work motivated us all. Each new day in the field might yield a noteworthy observation: a moose in a new canyon, a rare fleabane on a remote ridge. Each expanded our knowledge of this 2.2-million-acre Eden, all but hidden in the heart of Wyoming. There were nights I could barely sleep anticipating the next day's work of discovering and fitting the next piece of a complex puzzle.

PART II

Climb the mountains and get their good tidings. Nature's peace will flow into you as sunshine flows into trees. The winds will blow their own freshness into you and the storms their energy, while cares will drop off like autumn leaves.

—John Muir, *Our National Parks* (1901)

3

First Elk

Two weeks after arriving in Wyoming, I saw my first reservation elk from a Cessna 182 airplane. Dick Baldes cobbled together the dollars for six hours of flight time over winter ranges in the Wind River Range and Owl Creek Mountains. The new wildlife program's budget made no allowance for such niceties. For the remainder of fiscal year 1978 it was spartan.

Typically, wildlife surveys are flown during winter or early spring, when snow restricts big game to limited areas. As a further advantage, the snow background makes animals stand out like Winnebagos in a parking lot. Well, not quite that clearly, as it turns out.

Besides counting observed animals, biologists classify them by sex and by age—sometimes distinguishing just young of the year and adults, and sometimes including additional age classes depending upon the species and objectives of the survey. From these classifications, we determine sex ratios (expressed as adult males per 100 adult females) and recruitment (number of young per 100 adult females). Wildlife managers then estimate the size and makeup of populations, charting trends over time like financial planners gauging the past and future performance of an investment portfolio. Later in my career, I would use computer-based population models to do the math. In minutes they would churn out predictions of population growth and harvestable surplus that I had to painstakingly derive from simple calculator computations while at WRIR. Because this first flight occurred on May 20, animals were scattered. Males had shed their antlers already, and last year's calves, fawns, and lambs—now nearly one year old—were putting on post-winter growth, making them hard to distinguish from older animals. It was little more than an orientation flight—an opportunity for Dick to show me areas where he knew elk, mule deer, and bighorn sheep wintered.

Flying conditions can be iffy in spring. That day was clear but winds aloft were gusty, up to 40 knots as my stomach recalls. As the westerly flow tumbled down the eastern tilt of the Wind Rivers, the turbulence buffeted our single-engine, four-seat aircraft with bumps, lurches, and sudden drops in altitude that were downright unnerving. Compounding the effect, our pilot steeply banked and tightly circled each group of animals so we could count them. I felt my cheeks trying to peel off the underlying bone as a trickle of sweat traced my spine. Eyes glued to animals slipping by at 60 miles per hour, I often lost track of the horizon; the reference to which way is up. That is when the head grows light. Then the Cessna pitched onto its opposite side, and *hello vertigo!* Welcome to wildlife surveying.

Some sage wildlifers advise eating nothing before a flight. Others hazard the emptying of a well-filled stomach when air sickness strikes. Sitting behind the wing in the back seat, I had only an oblique view of the horizon ahead. Worse yet, each bump was exaggerated in the back compared to Dick's wing-side seat next to our pilot. Fortunately, I had been through this before.

By now I had logged over 200 hours in single-engine airplanes—Piper Cubs, Super Cubs, Citabrias, and Cessnas. During two seasonal appointments with the USFWS in

Elk surveys in Wind River Mountains from a Cessna 182, May 1978

Sheridan, Wyoming, in 1976 and 1977, I worked with five other biologists capturing and radio monitoring grouse, eagles, owls, hawks, pronghorns, white-tailed and mule deer, bobcats, and coyotes. We fitted hundreds of animals with radio transmitters embedded in collars and tiny backpacks, or glued to the base of tail feathers. I was the newcomer to the project and did whatever the others needed.

Several of my colleagues had no stomach for three hours of tight circles in cramped cockpits, so I became the designated radio-tracker. I learned the art of directing pilots to the "ping, ping, ping" cadence chiming in my headphones to pinpoint a signal. I saw the big picture of our study area spanning the Wyoming-Montana border and witnessed in "real time" how each species used the landscape. I savored the challenge of finding the nesting location of wayward sage-grouse or the summer retreat of a pronghorn 40 miles distant from its winter capture site. Each flight brought new surprises: another glimpse into the life of a deer I helped wrestle and radio-collar in a clover trap, or a sharp-tailed grouse I knew only by its unique radio frequency. It wasn't that I had become immune to the vagaries of low-level flying. But the feeling of being juggled like clothes in a washer was no longer novel, and was therefore less alarming. I didn't think I was going to crash and burn with every 100-foot plunge toward treetops so close I could sometimes count pine cones.

🦌 🦌 🦌 🦌 🦌

It was in July 1978, two months after beginning my new job, that I got my first look at a reservation elk from the ground. I was on a horse pack trip deep in the backcountry of the Wind River Range to inventory fish populations in several high mountain lakes. These fish sampling excursions, engineered by Dick Baldes, were eagerly anticipated by him and our station's other fisheries biologist, Reg Reisenbichler. I was excited to be invited on this first pack trip of the season, and my first ever. Our secretary Betty, a true horse lover, was left to tend the office. We would hear about that for several days upon our return.

Our destination was a cluster of lakes nestled at the headwaters of a Wind River tributary. In 1938, the tribes and the secretary of the interior set aside an 180,387-acre roadless area in the reservation's Wind River Range, one of the most scenic ranges in the lower 48 states. This was a visionary achievement, predating the U.S. Wilderness Act by 26 years. It permanently protected a pristine, glaciated landscape of subalpine meadows, virgin forests, crystalline streams, over 200 lakes, and peaks soaring above 12,000 feet. It also provided a haven for reservation wildlife.

Mid-morning, we arrived at the trailhead—more accurately, the place where the road became too steep and muddy to pull horse trailers any farther. We saddled up, cinched the Decker saddles onto the two pack animals, and loaded the canvas panniers. Loading a packhorse was as foreign to me as ballroom dancing, so I mostly observed and handed the others whatever they needed.

"Toss me those sleeping bags."

"That green stuff sack of clothes."

"Now the griddle. No, damn it! Don't toss that!"

"Now that picket."

"What?"

"The picket."

"Ummm."

"Over there. That steel pin with the long rope attached! Haven't you ever packed a horse before?"

I thought we covered that during my job interview, I mused.

After making sure fishing rod cases, sleeping pads, and other odd-shaped stuff were secured beneath the manti tarps, and wedging a trail saw and ax securely beneath lash ropes, we plopped onto the pickup tailgates for a snack before heading out. I passed around a bag of chocolate chip cookies, trying to make amends for not knowing what a picket was. When they came back to me, I took another and set the bag behind me in the pickup bed. Over our deliberations about whether we would be camping on snow tonight, a hideous coughing arose behind me. My designated saddle horse—a parrot-faced, barrel-bellied roan named Chippewa—was trying to extricate a plastic bag from his mouth. Moments before, that same bag had contained a dozen chocolate chip cookies.

"Git that away from him!" Dick chided.

I latched onto the horse's halter and wrenched the shredded, saliva-soaked bag from his mouth—*sans* cookies.

"That horse will eat anything! If he sees it and thinks it's food, he'll eat it," Dick chuckled.

"Can't be good for him," Reg chimed in.

As we joked about the parrot-faced food Hoover, Dick noticed a bag flapping on the ground next to the truck he and Reg were perched on. There lay a half-empty five-pound bag of red potatoes. No one had noticed them missing during the bustle of packing.

"By God, I told you he'd eat anything!" Dick howled.

We all made a pass around the vehicles seeking half-devoured produce, steaks, or toothpaste scattered in the grass or stomped in the mud. Finding none, we assumed "Garbage Gut," as he became known, had destroyed the evidence. We would learn what else he had dispatched by its absence when we unpacked the panniers at camp. As it turned out, Chippewa was just getting warmed up with surprises.

We locked the trucks and mounted up. Dick headed out, followed by Reg, each leading a packhorse. I had been on a horse just enough times to say I had ridden, so the greenhorn brought up the rear. Not 50 yards from the vehicles was a large snow patch maybe two feet deep in the middle. Dick and Reg circled around, but my horse just headed straight for it. I reined him left but he wouldn't be turned. In a moment, he lumbered onto the still frozen patch, dropped to his knees, and without hesitation rolled onto his side.

Man, I didn't see that coming! Just before my right leg was trapped beneath, I pitched myself free of the saddle. Sprawled on the snow, the horse rolled beside me, wearing a self-satisfied expression.

As the laughter up ahead subsided, I gathered myself, coerced the beast back to his feet, and led him off the snow. Whether he had gotten vapor locked from the spuds or had just wallowed for pleasure, I can't say. I can say I was both wary and pissed at the horse. Without looking at my companions, I wiped snow from the saddle seat with a bandanna, straightened the saddle bags, and checked the cinch. Muttering a few choice words at which the others cackled, I remounted and began the trip again, this time with a very short rein on Garbage Gut!

Backcountry trails on the WRIR were sparsely traveled, largely because non-Indians could only "trespass" on reservation lands for the purpose of fishing in designated areas. Nowadays, most Shoshones and Arapahos didn't venture into the Wind River backcountry. Thus, riding the trails required clearing the trails.

We were obviously the first on this ten-mile stretch of rock, mud, and blowdown this season, which we carped about and yet enjoyed. Early July is late spring in the Rockies. The trail repeatedly vanished beneath crusted snow as we passed the 9,000-foot contour line. Our horses would hesitate and then plunge into the snow with further prodding. When they floundered and stood miserably, we would dismount and exhort them through. The afternoon sun thawed the snow, which balled beneath the horses' feet, clung in frozen clods to their fetlocks, and soaked our jeans. By the time we reached our destination, we were leading the horses more than riding them.

We approached the first of a chain of lakes, where Dick chose to make camp. It lay veiled by ranks of dark firs and spruce—like soldiers standing guard at their posts. Half encircling the lake's far shore rose rival ramparts, ground smooth and sheer by Pleistocene ice. We topped a final rise holding the lake's dark waters in check. The glassy reflections of talus and peaks were hypnotic beneath white cumulus ships sailing an azure sky. This was one of those places that reassured my heart I had chosen the right course for my life.

The snow lay in scattered patches, melded to last year's mat of sedges and grouse whortleberry. Beneath breaks in the forest canopy the snow had surrendered to the sun's radiant heat. To a wall-tent-sized spot of bare ground we reined the horses. Then the unpacking began.

But first I led Chippewa toward the driftwood-littered shore. I took a moment to just soak in the surroundings and allow my knees to recover from stirrup strain. On the far ridge to my left, the one place gentle enough to cultivate an alpine meadow, a thin line of specks angled toward the skyline. Raising the binoculars slung around my neck, I saw seven rusty bodies trimmed in chocolate manes and ginger rumps melt into the sky. I remember this image perfectly, for perfect it was. These were the first reservation elk I had seen, not from the cramped vantage point of a droning Cessna, but standing nearly 10,000 feet above sea level and 10 miles deep in the Wind River Range—beside my trusty mount, Chippewa. The horse tossed his head, tugging the reins through my gloved hand. Back to reality. Garbage Gut sensed dinner being unpacked.

I spotted just two other groups of elk during our five-day excursion. The first bunch dissolved at a lope as soon as I saw them and they saw me. At a half mile distance, was I a threat? I spied the other bunch of 20 grazing a meadow tucked in an adjacent tributary canyon. These were wild elk, like part of the mountains, born and bred to live sovereign and free.

🦌 🦌 🦌 🦌 🦌

Taking a well-deserved respite from reeling in big-spotted Yellowstone cutthroats, I climbed to the summit of Roberts Mountain. My destination was the place where I had glimpsed the column of elk two days before. I told the others I wanted to look around for elk, mountain sheep, or whatever I might find up there.

From a distance, alpine ridges look smooth and verdant with only random boulders suggesting tough footing. Climb there, and that distant garden transforms to a cobble of lichen-encrusted rock with vascular plants squeezed wherever the slightest suggestion of soil consents. I picked

Roberts Mountain, Wind River Mountains, July 1978

my way south across the broad ridgetop, marveling at how quickly blooms of blue, yellow, pink, red, and white burst from plants just freed from eight months of snow. All hugged the ground in this wind-racked environment—sheep fescue, alpine avens, moss campion, and a spray of other "cushion plants." A marmot whistled somewhere in the distance, recently awakened from months of hibernation. Unlike her eastern cousin, Punxsutawney Phil, she was buried beneath several feet of snow on Groundhog Day. Emerging to a colossal salad bowl, she would mate, raise a litter of three to eight, and increase her body mass by 30 to 50 percent over the next few weeks. She would rival a fat house cat's weight, then repeat her Rip Van Winkle act in August.

A palm-sized pika, or "rock rabbit," North America's diminutive member of the hare family, scolded from a rubble pile to my right.

"E-e-e-ek. E-e-e-ek."

Unlike the marmot, this solitary alpine farmer does not hibernate. Instead, she readies for winter by busying herself all summer harvesting crops. She stores them as winter provisions in hay piles beneath rock ledges and boulders. The cutting and stacking go unabated as she raises two litters each summer. Long predating the farmer's root cellar and stockman's haystack, the pika was among the first to hoard food for winter. Studies in Colorado revealed that not all vegetation in hay piles is used during the winter. Hence, the pika is a bet-hedger, storing surplus food as insurance against an unusually long winter.[1] Because of her small size and high metabolic rate, the pika carefully judges the nutrition of grasses and forbs, harvesting those that afford the greatest caloric return. Both males and females aggressively defend territories surrounding their stored provisions. The assertive survive winter; the timid may not.

Traveling causeways below the boulders, a pika unpredictably pops up like a submarine's periscope to scan for danger. This one gave me a good look, then disappeared into a maze of jumbled granite. The pika joins a host of other obligate alpine animal and plant species bound to the West's islands-in-the-sky wilderness. As the highest mountain range in Wyoming, the reservation's 40 miles of Wind River mountaintops host an assemblage of the continent's most pristine biodiversity.

Far removed from most pollution, habitat ruin, and exploitation at the hand of man, the pika would seem the least likely of species to be considered for the U.S. endangered species list. Yet, three decades after this encounter on Roberts Mountain—when I was as naïve as the pika about its grim future—that is the case. On a warming planet this five-ounce boulder bunny's cool mountain habitat is in retreat. Adapted to cold alpine climates, pikas are subject to heat stress in warm temperatures. They can perish when exposed to temperatures of 78 degrees or more for just a few hours. Warmer temperatures may also limit food supplies and reduce the snowpack required to insulate them in winter. More than a third of lower-elevation pika populations in the Great Basin mountain ranges of Nevada and Oregon have disappeared in the past century. Those that remain live an average of 900 feet farther upslope to escape the summer heat. Studies in California corroborate the pika's retreat.

Climate experts estimate temperatures in the western U. S. will increase during this century more than twice as much as they did last century. This is expected to eliminate the pika, and perhaps other heat-intolerant alpine species, from large regions of the American West. In a climate out of control, the highest peaks of the Wind Rivers and other lofty ranges may be the last refuge of alpine species. Adaptation simply cannot keep pace with the changing climate. This bodes especially grimly for species confined to island enclaves and unable to migrate. As a result, the USFWS formally acknowledged in May 2009 that global warming may threaten the pika and undertook a species status review.[2] This pint-sized bellwether seemed destined to be the first mammal designated a threatened species due to global warming. However, the USFWS ruled that although the American pika is potentially vulnerable to the impacts of climate change, the best available scientific information indicates that pikas will be able to survive despite higher temperatures.[3]

Effects of our planet's warming on the persistence of life are greatest in the polar latitudes and at the highest altitudes where temperatures are rising most rapidly. Polar bears, penguins, pikas, and mountain goats are at rising risk of extinction—something only our species, not rock rabbits, can rectify.

I scanned among the heath and asters for fresh elk droppings—not the barrel-shaped pellets produced when forage is cured, but soft, amorphous clumps of pungent dung. The high water content and digestibility of early plant growth fail to form perfect pellets. *There!* Poking a chocolate cluster with a heather stem, I exposed its moist green interior. The smell of sweet barnyard perfume rose from the dung. This was fresh, maybe a day or two old. In a nearby puddle of decomposed granite, a 4.5 x 3 inch cloven impression registered the passing of a full-grown elk.

I crested the summit and began descending when 20 elk appeared beyond the far shoulder of the ridge. Several cows were bedded beneath a scatter of conifers. Others casually cropped sprouting delicacies as their calves frolicked nearby. Some calves sprinted and dodged haphazardly as though afterburners had been ignited. Two others butted heads in mock combat. Their white spots projected on a rusty orange backdrop, each of these awash in an emerald meadow.

Safely unnoticed on the mountain high above, I sat and watched. Like the marmot and pika, the nursing cow elk were recovering body condition depleted by the rigors of winter and pregnancy. Moreover, energy demands on each mother on her first day of lactation were much greater than required on her last day of gestation, when *in utero* energy transfer is highly efficient. Even though milk is nearly 100 percent digestible, 40 percent of the energy is lost from mother to nursing calf. It is lost through conversion of milk into tissue, and through the increased energy expended by a romping calf compared to a womb-bound fetus. The calves I watched pumping their mothers' udders for another meal were gaining two pounds of body weight per day. This cost of lactation to these nursing mothers peaked about 48 days postpartum—just about now.

Nursing calves were increasingly switching to a diet of plants to meet their growing energy needs. In order to both maximize milk production and aid the conversion of growing calves to an herbivorous diet, selection has favored synchronizing births during late May through mid-June, when plant phenology and growth are exponentially increasing. Born too early in spring, a calf risks malnutrition from inadequate milk, as well as thermal stress from late spring snowstorms. Born too late, a calf faces too few days of grazing nutritious plants, and may not grow large enough to survive the upcoming winter. It is a tightrope natural selection has precisely walked.

I felt a chill and saw that a fierce, tumbling bank of clouds had swallowed the afternoon sun. They rushed from the west. The air temperature plunged. Before tucking the binoculars inside my parka, I glassed the little herd of elk one last time. They seemed oblivious to the approaching storm. Starting back up the ridge, thunder rumbled in the distance, then again closer. I began making tracks.

As I crested the ridge, the first wind-driven raindrops pelted me. I could see the nearest cover, a stand of whitebark pines still a quarter mile ahead. They looked to be scant protection. Lightning streaked and the trailing thunder rapidly gained with each flash. Minutes later my arms began tingling. Pulling up my sleeves, I was alarmed to see the hair twisting into tiny punk spikes. I felt my hat rise on its own. Static electricity engulfed me. A bolt blasted the ridge off to my left and belched an acrid electric smell.

That was enough! There was no escape. Standing, I was the tallest object on the ridge, though I had seldom felt punier. I ditched my binoculars and metal-framed daypack beneath a jutting flat rock, and sprinted for shelter beside the only boulder within 100 yards.

As quickly as it advanced, the storm bore east. Within 15 minutes, I was standing in brilliant sunlight, transfixed

Crowheart Butte, 1978

by a dazzling double rainbow awash in drizzle from a cloudless sky. It was like a scene from the Wizard of Oz. Far to the east, lightning challenged Crowheart Butte, a sandstone pyramid rising 700 feet above the surrounding prairie. Near there in 1866, Chief Washakie bested his rival Crow chieftain, Big Robber, and cut out his heart. Washakie's triumph secured the Shoshones' supremacy over hunting grounds in the Wind River Basin.

🦌 🦌 🦌 🦌 🦌

I would watch elk on other days in the Wind River Range and Owl Creek Mountains, but none would surpass this one. It felt like a private viewing of a rare masterpiece and rekindled the special affection I discovered for elk during 1973. That summer I worked on the Lolo National Forest in western Montana for Gary Halvorson, a tall, sandy-haired, soft-spoken man of 40. As on WRIR, Montana's elk spent summer and fall in remote high country, including roadless tracts of big trees coveted by the Lolo Forest's ambitious foresters.

In contrast to the Lolo's well-funded and ample silviculture staff, Gary was the only wildlife biologist on the Lolo, the second largest national forest in the lower 48 states at the time. More than once I watched him return still seething and dispirited from an interdisciplinary team meeting where more pristine forest was proposed for sawlogs. It wasn't just that Gary was forever playing defense, pressed to justify why habitat for these or those species in this or that watershed should not be clear-cut. Ostensibly, the outcome was often a foregone conclusion as he faced ten to twenty foresters hell-bent on "getting out the cut."

It worked like this: regional foresters presented national forests with annual timber targets, like generals issuing orders to Marines. Targets were measured in millions of board feet. Meeting those targets figured into performance evaluations of staff and supervisors—and perhaps promotion potential. Gary seemed to me a lone voice in the wilderness. He was the gloomy Gus reacting to the current crisis. He alone spoke for the wildlife that had no voice. These were the days before Environmental Impact Statements or legal challenges to timber sales were in vogue. I felt deeply sympathetic. As a lowly field technician, I was in no position to help, but I resolutely pledged that someday I would shape such heady decisions. For now, though, nurturing my skills and building a résumé would have to do.

In 1973, Gary went on the offensive. To proactively identify important habitats and better defend wildlife needs, he initiated an inventory of elk ranges in five planning units of the Lolo. These were large swaths of *de facto* wilderness, each greening tens of thousands of acres. Gary hired my coworker Tad and me to map all significant elk habitat features. For 16 weeks we backpacked where mountains met sky, camped beneath stars, wolfed heaps of freeze-dried food, enjoyed breathtaking landscapes, saw loads of wildlife, rarely saw people, and lost what little baby fat either of us had. It was a blast!

In some respects it was a treasure hunt, always into a new corner of the forest we had never seen before. Along the way we found old trapper cabins—the roofs often collapsed—littered with old cans, rusted tools, and telltale newsprint. The hand-hewn logs and sod roofs spoke of a time when men were men, beaver grew scarce, and buried riches proved elusive. On occasion, a rocky fire ring overgrown by vegetation, a sun-bleached elk skull with the antlers hacked off, or a well-worn horseshoe showed the passage of unknown souls. At other times, I felt sure no other human had ever set foot where my boots trod. Such places included mucky alder thickets, fir forests where I tiptoed along deadfall high above the forest floor, and slopes I traversed that were "steeper than a cow's face" in mountainfolk speak.

We were strictly on our own, scouring prime black bear and sometimes grizzly habitat. We carried only a two-way radio as a safety net. Each morning at eight, we checked in with the Lolo Forest's dispatcher. Two missed calls and a search party would be organized to scout our remains.

We split up each morning to explore the dendritic headwaters of stream courses, moving in stealth mode like hunters on a hot trail. This increased our chances of stealing upon elk and spying on their daily lives. Bushwhacking miles each day, we diligently plotted elk bedding areas, wallows, and watering holes. When time permitted, I followed and mapped the routes of heavily trodden trails. They often led from one important biological feature to another. Traipsing from one small drainage to the next, I would creep to thin ridgetops, never knowing what I would be treated to on the other side—a group of calves prancing across a garden of lupine, mule's ears, and geranium under their mothers' watchful eyes; a bull thrashing a spindly fir to strip the remaining velvet from his hardening antlers.

Sounds and smells were telltale signals of nearby elk: a sharp bark from a cow to warn herdmates of scented danger; the languid talk of cows and calves winding through a the parchment trunks of an aspen grove; the "mewings" of calves floating softly across a basin on the breeze. And by late August, the bugling of bulls in rut erupting in earnest. Not just the occasional warm-up I strained to locate earlier in summer. No, these were strident shrieks, bellows, and grunts that announced their readiness for love and war. To me it proclaimed, "This is wilderness, no logging allowed."

One mid-summer day, I believed I had discovered elk utopia in a drainage named Big Sunday Creek. To my eye, this glacier-gouged, U-shaped canyon epitomized the best of summer range. Spruce and fir forest was liberally quilted with flowered meadows, aspen groves, cobalt ponds, and gushing water. This was where I would spend summer if I were an elk, or a human for that matter. As I worked my way down the initial watercourse, the knot of trees ahead erupted with running animals. How many, I couldn't tell. I glimpsed not even one. But throughout the two days I spent mapping Big Sunday, the musk of elk greeted my human's weak sense of smell again and again. I did see elk eventually. Their rusty orange coats and glistening long necks sliced vertically among stolid spruce trunks. But the abundance of droppings, trails, and bed sites pressed into grass and sedge conveyed the larger story. This should tell all, to biologist and forester alike. Reserve this basin for elk and their wild kin, not for sawlogs and haul roads and forever lost solitude.

As I visualized invading chainsaws and bulldozers, I was struck with admiration for Gary's dogged efforts. He would much prefer being up here to arguing against a conference room of timber beasts. While Tad and I supplied him bullets, only Gary's skirmishes could guard these sights and sounds of elk for future enjoyment by others. His job was where the rubber met the road.

As summer wore on, I found myself wanting to know more about these magnificent creatures. Moreover, would the information I collected sway decisions forever affecting their mountain retreat? My hunger to learn and protect the needs of their harmonious community of life spurred me like nothing before. That summer living among wild, freely roaming elk solidified the purpose of my life's work.

Now I felt thankful and privileged working here in the Wind River tribes' homeland. What income they had was derived from their land. Yet time and again they had chosen the spiritual wealth of transcendent wild Nature over monetary gain. As one testament, in 1971 they shut down commercial logging of their mountain ranges. Tribal members had decried their land being despoiled and favored reserving it as elk, deer, and bighorn domain.

🦌 🦌 🦌 🦌 🦌

Sightings of elk during future forays into the Wind Rivers and Owl Creeks were always exciting but tryingly uncommon, at least until I better learned their haunts and habits. How differently they behaved from the elk I watched on weekend trips to Yellowstone National Park—elk that had little to fear from humans. WRIR elk were either exceedingly few or exceptionally shy. A bird's eye view confirmed the latter.

Surveys from light aircraft provided the bulk of my data to assess populations of elk and other big game. I flew as frequently as budgets permitted, often accompanied by a tribal game warden or another of our station's staff. Larry Hastings had piloted Dick and me over the Wind Rivers and Owl Creeks in his Cessna 182 during my inaugural flight in May 1978. He was an affable 30-something local, who could have passed for a California surfer with his mop of streaked blond hair. He remained my sole fixed-wing pilot on the WRIR. Don't mess with a good thing when it comes to technical, low-level flying.

Larry was a pro at negotiating the rugged terrain and Venturi effect of winds funneling through canyons at high velocities. Still, I found sorting groups of elk, deer, and sheep (calves from cows, fawns from does, ewes from young rams) problematic from the Cessna. Stall speeds in excess of 45 knots prevented Larry from getting me more than a quick look before the plane zoomed by and our quarry headed for the nearest cover. The fixed-wing was fine for counting animals. But on these surveys I classified only 68 percent of the elk we spotted. Doubt remained whether the classifications were representative of all the elk we saw. A biased sample could be misleading, but there was no way to determine bias without a reliable classification of all observed elk.

Federal funding was insufficient to charter a helicopter, indisputably a more capable but costly survey craft. So I approached the Tribal Fish and Game Committee and subsequently the Joint Business Council. I walked them through our survey methods and stressed that I wanted high confidence in the results I gave them. Convinced of the need, the councilmen promptly diverted $14,900 from other tribal programs to a special wildlife account. They earmarked $10,000 for flight time, and the remaining $4,900 to begin inventorying vegetation on big game winter ranges. These were not wealthy people and my requests were substantial. Still new to the job, and to them, I was elated by their broad support. What a pleasure working with a government of only two layers, especially when members of the first layer also served on the second.

Over the ensuing three winters, half my survey hours were in the Cessna, and half in three-seat, turbocharged Bell and Hiller helicopters. From the choppers, I classified 85 percent of the elk observed, cutting the number of unclassified animals more than 50 percent.[4] Most of those that I could not separate as spike bulls (yearling males sporting unbranched antlers), adult bulls (branch-antlered males), cows, or calves were spotted in woodlands, or fled there at the oncoming "thwack, thwack, thwack" of the rotor blades. Like labeling Republicans and Democrats in the synthetic fog of a lobbyist-sponsored mixer, classification reliability was poor in forest cover. I had to assume the makeup of the unclassified elk was similar to the other 85 percent.

Reservation elk were more migratory than other big game except pronghorns. Elk moved downslope from their summer ranges both on and off the reservation in the Wind River Range as snow accumulated above 8,500 feet. The elk of the Owl Creek Mountains were largely seasonal residents of WRIR. Most spent summer and fall in the Absaroka Mountains, mingling with elk resident on national forest lands to the west. They migrated as many as 20 air miles eastward to the reservation during November and December.

I counted the largest numbers of elk during bouts of severe weather—both numbing cold snaps and immediately after heavy snowfalls. The winter of 1978–79 was brutal, far colder and snowier than the following three winters. Accordingly, I observed more elk that first winter. My first flight in the Cessna on November 15, 1978, followed a record storm. Three feet of snow blanketed the Lander airport; five feet buried higher elevations in the mountains. Old-timers commented, "I ain't seen this much snow since granddad was alive."

November through January, temperatures registered the coldest on record for that three-month period. Above normal snowfall continued. I watched the elements take their toll as in no other year of my tenure, and in few others before and none since.

Six aerial surveys of the Winds and four of the Owl Creeks provided a timeline to assess population change and the survival of calves, the most vulnerable animals due to their small size and limited fat reserves. The ratio of calves per 100 cow elk in the Winds declined from 43:100 in November to 17:100 on late February and March flights—a 60 percent loss of calves. All but 10 percent of the loss had occurred by mid-January. Because I derived the loss from a change in ratios, if cows also died, then the calf loss was even more precipitous. In the Owl Creeks, mortality of calves was similar at 62 percent.[4] Winter mortality of this magnitude had rarely been reported elsewhere. However, few elk herds are repeatedly surveyed during winter to quantify such losses.

By comparison, 13 percent and 18 percent of calves (Owl Creeks and Wind Rivers, respectively) perished during the more normal winter of 1979–80. As Charles Darwin wrote in *On the Origin of Species* nearly 150 years ago, "all nature is at war," and "the struggle very often falls greatest on the young, but fall it must sometime in the life of each individual or more commonly at intervals on successive generations."[5] I had witnessed such a generation fall. The weak were weeded out like chaff threshed from grain. Those left were surely the most physically fit among all calves of their cohort.

Remarkably, my highest counts of elk in the Owl Creeks were nearly identical those first two winters, 434 and 447. This may have signaled a shift in distribution. More elk from national forest lands may have migrated to the less snow-prone WRIR in 1980—possibly compelled by ghosts of the previous winter.

In the Wind Rivers, however, I counted 1,584 elk in early February 1979, but only 907 a year later.[4] Did this reflect a real decline in that population? Were the two flights budgeted during the winter of 1979–80 both flown on ill-timed survey days? Or were elk just more difficult to find during the milder winter of 1979–80 in the more heavily timbered Wind Rivers compared to the grass- and sage-dominated Owl Creeks? Possibly some of each. These are matters that vex biologists.

I concluded from the surveys that 2,000 elk wintered in the reservation's Wind River Range and 550 in the Owl Creek Mountains. This incorporated a 20 percent "fudge factor," what wildlife scientists term observability bias. I boosted my highest counts by 20 percent to account for elk likely not seen on each winter range. Why 20 percent? The distribution of elk pellet groups my inventory crews and I sampled showed that just 20 percent occurred in forests versus more observable habitats. An 80 percent sampling efficiency also concurred with Dick Knight's findings for Montana's Sun River elk herd. On similar winter range, Knight calculated an 83 percent sampling efficiency for aerial surveys, based on the proportion of marked animals he observed from a total sample that had been adorned with brightly colored neck bands. Short of marking every one, a definitive total is rarely known for any wild population.[6]

Given that all estimates of naturally occurring populations are imperfect, though some are less imperfect than others, I explained the numbers to the Joint Business Council in a slide presentation. They were collectively pleased but surprised, expecting far fewer elk to be found. One Shoshone councilman fond of jabbing me asked, "Did you count them all twice?"

🦌 🦌 🦌 🦌 🦌

So it went with the mule deer, bighorn sheep, and moose I surveyed. I developed estimates of population sizes, annual productivity, habitat selection, and best guesses of harvests by hunters (no reporting system existed). I delineated seasonal ranges of each species and locations where translocating bighorns could reestablish lost herds. I even measured summer and winter diets from fecal pellets my co-workers and I methodically bagged like gathered wild raisins. In a laboratory at Colorado State University, white-coated technicians examined the feces. Through microscopes they identified plant composition by the characteristic cell wall structure of each genus or category of plants (who said science isn't fun?). *Voilà*—5.9 percent *Phoradendron*, 3.5 percent *Mahonia*, etc.

Big sagebrush amounted to 6 to 18 percent of bighorn sheep diets across winter ranges I sampled in the Wind Rivers. Studies elsewhere showed sagebrush of varying importance in winter diets of this largely grass-grazing species. In the Wind River Range, reservation bighorns were largely confined to the slopes of Dinwoody and Red creeks in winter. Fewer than 100 wintered there, although their range was contiguous with additional sheep on the Shoshone National Forest to the northwest.

Because they shared wildlife ranges, I also quantified diets of free-ranging horses and cattle. Their winter and summer diets (horses grazed game ranges year-round) exceeded 95 percent grasses. Among WRIR big game, elk (58 percent) and bighorns (53 to 68 percent) were the primary grass consumers. I concluded that livestock posed more foraging competition to the struggling bighorn population than did other wildlife. Notably, the sliver of bighorn habitat in Red Creek was a regular hangout for two dozen feral horses, but was rarely grazed by elk.[7,8]

On reflection, the diet data were "gee whiz" information. It is not as though one could change an elk or deer's food preferences to accommodate what grew most abundantly. Theoretically, beneficial land management practices could favor the preferred food plants of wildlife. While providing the tribes this information had no downside, no monumental management decisions arose from how much

Artemesia tridentata (big sagebrush) reservation mule deer ate in winter. From a holistic perspective, however, I find it instructive that the big sagebrush so critical to the survival of disappearing populations of North America's sage-grouse is likewise a diet staple of mule deer, a common and much cherished game species. No species I'm aware of is more closely coupled with sagelands than the sage-grouse. Lewis and Clark are credited with "discovering" what Clark called the "Cock of the Plains." Of course, aboriginal people from the Great Plains of southern Canada to the basins abutting the Cascades and Sierras lived among and hunted these birds for generations.

In the predawn light of April, I would visit sage-grouse leks. At these strutting grounds, I counted how many flamboyant males gathered for the morning courtship dance. By filling and deflating paired, olive-colored air sacs, males produced popping, bubbling sounds and flared lancelike tail feathers, ever trying to impress. As biologists do elsewhere, I could most successfully evaluate WRIR sage-grouse numbers by finding where they assembled for their breeding season hoedowns.

Some tribal members, I was told, remained fond of roasted sage chickens, their name in the local vernacular. Given year-round hunting, the birds would be particularly vulnerable when the seven-pound cocks gathered to strut their stuff for the hens each spring. To my relief, I found no one laying waste to dancing grouse during my surveys of leks.

But beyond protection from overexploitation, paramount to all species survival is ample habitat, the currency of wildlife. Sage-grouse need sagebrush for food and cover, just as black-footed ferrets—North America's most endangered mammal—need prairie dogs for nourishment and prairie dog burrows for refuge from watchful soaring eagles.

Those empty-looking expanses of sagebrush symbolizing the Intermountain West support their own rich community of obligate fauna. Sage thrashers, golden eagles, jackrabbits, pronghorns, and prairie dogs might fade away without sagebrush, just as ferrets would vanish without prairie dogs. Tragically, robust sagebrush habitats have fallen prey to energy development, subdivision, agriculture, and other perils across the West. Though sagebrush was once reviled and despoiled as a mere competitor of grasses grazed by livestock, we now understand its ecological importance and should value sagelands accordingly.

Fortuitously, the bighorn sheep fecal pellets we bagged served double duty. Bighorn populations in Colorado, Idaho, and the Canadian Rockies have suffered high mortality from what is termed "lungworm-pneumonia complex." As the name implies, lungworms—which look like angel-hair pasta—infect the respiratory system of bighorns. Severe infestations obstruct and irritate tissues, predisposing animals to life-threatening pneumonia. Lambs are most vulnerable to pneumonia within a herd.

Sheep shed larvae of adult worms in their feces. Fecal larvae counts mirror the severity of parasite infections. Large loads of larvae per gram of feces correlate with excessive ungulate numbers on sheep ranges, range deterioration, and nutritional stress—bad news for bighorns. Fortunately, Dr. Robert Bergstrom at the Wyoming State Veterinary Lab found that the lightly stocked reservation habitats yielded very low larval counts.[8]

By contrast, northwest of the reservation at Whiskey Mountain, 1,000 bighorns—Wyoming's largest and densest population—have chronically high lungworm loads and suffer episodes of disease.[9] Forage use there averaged a remarkable 81 to 87 percent of what grew during the 1970s, compared to the 19 percent use I measured on reservation winter ranges.[8] My job was to advise the Shoshone and Arapaho tribes how to expand big game numbers. But I would not advocate unsustainable numbers that would damage habitats and the health of herds. Any short-term advantage of overburdened habitat is ultimately lost, like the broken-down springs and shocks of a sorely overloaded truck.

4

Mountains and Sky

Long weekends usually found me deep in the Wind Rivers. Backpacking had infused my blood during my graduate studies of mountain goats and summer employment shadowing elk. Buoyant self-sufficiency and escape from civilization's confinement quickened my boots and my heart, as did launching a canoe for a weekend float. Perhaps it smacked of the self-determination and wonder that drove fortune-seeking easterners to launch Conestoga wagons crammed with all earthly possessions into the western frontier. With horses I could have traveled farther and certainly in greater luxury. But pack trips require more attention—including care and feeding of the horses—and limit travel to passable trails and gentler terrain. Only technical skills and good sense limit mountaineering routes.

The reservation backcountry, particularly an 180,000-acre designated roadless area in the Wind Rivers, was virtually vacant of humans. With my job came license to roam these pristine landscapes where other non-Indians could not. Cradled in hanging canyons, lakes named Moccasin and Moraine, Solitude and Steamboat, Wykee and Heebeecheeche were shadowed by granite ramparts ascending to summits named Windy and Wolverine, Shoshone and Dinwoody, Lizardhead and others as impressive, but unnamed on U.S. Geological Survey maps.

During one three-day weekend, I camped at a remote lake once stocked with golden trout. Goldens are not the most challenging or biggest trout; those inhabiting streams usually max out at a pound or so. Bigger fish grow in alpine lakes, with the largest recorded—up to 11 pounds—from lakes in the Wind River Range. But the species' most noteworthy allure to anglers is its distinctive coloration.

Intermittently a few of this lake's fish cruised within casting range of my seven-weight graphite. Yet my fur and feather offerings failed to ring the dinner bell. Finally, in the outlet stream's fast water I hooked a couple of two-pounders. These may have been spawners, judging by their blazing golden flanks and vibrant red stripes. The fish danced as rambunctiously as the silver cascade that grew them, then flipped and splashed into my net.

Golden trout are not native to Wyoming. Ancestors of these fish were transplanted from streams on the Kern River Plateau of California's southern Sierras. For 2,000 years, the Tubatulabal Indians caught them in rock corrals constructed in the Kern River and its tributary streams. *Tubatulabal* is a Uto-Aztecan word meaning "pine nut eaters." Pine nuts and acorns (ground with mortar and pestle) and fish were their diet staples. Separated by over 700 airline miles from their distant Shoshone relations in the Wind River country, who spoke a Uto-Aztecan language too, the Tubatulabal also hunted rabbit, deer, bighorn sheep, and Tule elk—the smallest of North America's four extant elk subspecies. The culture of these early Americans inhabiting the plateau draining Mount Whitney's southern flank linked tribes of the Great Basin and Rocky Mountains to those of California's mountains and central valleys. Fish and bison, pine nut eaters and sheep eaters, Tubatulabal and eastern Shoshone; these ecologies and cultures were bound across time and space by a common ancestry depicted in petroglyphs on varnished Jurassic sandstone.

Six years after leaving WRIR, I was invited to the Sequoia National Forest to help develop a conservation strategy for its vanishing aspen groves. The location was the Kern River Plateau. My hosts and I considered how suppression of wildfires and the invasion of cattle and summer homes had altered the landscape. In a log cabin surrounded by towering Jeffrey pine and white fir, we deliberated by incandescent light how a more natural ecology could be restored.

On my last afternoon, I walked the bank of a gliding brook and searched for remnants of rock fish traps. Colorful small trout darted across likewise golden gravels, but I glimpsed no vestiges of their prehistoric captors.

🦌 🦌 🦌 🦌 🦌

On those first Cessna flights in spring 1978, I spotted a cluster of cow elk in Bull Lake Canyon. They grazed a north-facing slope where the mountainside had once given way to gravity and slumped into undulating folds. The result was a complex topography congested with pockets of aspen, mountain maple, and alder interspersed with grassy benches and swells of gray granite talus, all bounded by pines and fir. Everything an elk could need was there—food, cover, and water gurgling from fissures underfoot.

I planned a three-day trip during early June 1980—the peak of elk calving season—to see the place up close. After an eight-mile boat trip to the head of Bull Lake, I hoisted my backpack and hiked another two miles to a flowered bank of Bull Lake Creek. It was the kind of spot meant to be slept in.

A year before, I had camped beside this snowmelt-swollen stream with Tom Lemke, a fellow graduate student from the University of Montana. Tom was completing a Ph.D. program at Rutgers University and desperately needed some mountain R & R. Nowadays a game biologist for the Montana Fish, Wildlife and Parks Department,

Tom oversees management of Yellowstone National Park's northern elk herd when they migrate outside the park. Just upstream from camp, we had watched a newborn mule deer brave the swirling current to reach its mother pacing the far bank. The fawn's gangly legs pumped frantically as the current flushed it downstream, then into an eddy where it clambered onto the far bank. Trembling and wet, the fawn collapsed in its newfound safe haven on a lush bed of buttercup blossoms. The mother nuzzled and licked her little one, abruptly trotted off, then paused to glance back as if to say, "Hey junior, let's get cracking!" The fawn in turn must have wondered, "Why couldn't we have just stayed on the stream's other side?"

Tom and I had spotted a handful of cows and calves on that same north-facing slope where I circled above elk in 1978. These two data points suggested this could be a traditional calving ground. Now, a year later, I glassed the mountainside trying to boost the sample size. Nothing materialized in the lengthening evening shadows. I planned to investigate more closely next morning.

As I gulped instant oatmeal and coffee, first light illuminated the banded limestone opposite the elk slopes. Those cliffs were the same that Wyoming's nongame biologist, Bob Oakleaf, and I would scour by helicopter for peregrine falcons a week later. We would find great potential nest sites and lots of ducks and other yummy birds to hunt at nearby Bull Lake, but no falcons.

Alone now in 1980, I set out as a mayfly hatch erupted from the creek's milky green waters. In the warmth of June sunlight they fluttered around me like gusts of ripened cottonwood down. My daypack was heavy with water, lunch, notepad, cameras, and rain gear. Neither rain nor snow was forecast, but only a fool risks foul weather in the Rockies.

With no trails to follow, it took most of an hour to reach the first of the slope's grassy benches. This was bushwhacking at its finest through brush and boulders; pure ankle-twisting, knee-bashing, pants-shredding bliss. I decided to skirt the series of small benches above, where elk would most likely be grazing in the cool of morning. Staying downwind, I wanted to avoid disclosing my presence so that I could spot any calves that might be nursing or bedded near their mothers—assuming there were elk to be seen. Again, none were visible from camp at dawn.

I climbed higher through a tangle of old-growth fir forest until the mountain straightened further and I found myself above the slumps. I picked my way through the talus and undergrowth toward a high spot offering a good vantage point. After a respite to listen for elk calling to one another, I soon grew impatient and began exploring the prime benches just below.

The meadows were lush with spring's new growth and studded with clumps of black manure. The pungent barnyard scent of elk lay heavy on the air. I hiked with binoculars strapped around my neck, but my 35-millimeter Nikon was buried in the daypack. Sensing that times like this create Kodak moments, I slipped off the pack straps and knelt to mount the camera on my tripod. I had learned this lesson stalking mountain goats in Montana's Bitterroot Mountains. Shouldering a tripod-mounted camera could mean the difference between an opportunistic snapshot and no photograph at all.

As the camera snapped into position, I heard or sensed something over my left shoulder. Into view materialized a broad head, stout neck, and thick-muscled body. I was face-to-face with the biggest black bear I had seen since my summer among elk on the Lolo National Forest in 1973. He was the color of dark coffee, with a rusty muzzle and piercing ocher eyes. Round as a whiskey barrel, he stood broadside just 15 yards away. He cocked his nose toward me and scented to detect if I was danger or food. Acutely hoping to

The black bear in Bull Lake Canyon, June 1980

avoid startling him, I spread the tripod's legs deliberately in front of me. Hair bristled on my neck when I focused the camera's lens and found the bear half filling the viewfinder and staring back at me. The first "clack" of the shutter sent him bustling with a hoarse "Wuff!" into the forest, leaving me with that lone confirmation on Kodachrome that this had ever happened.

That remains one of my all-time favorite photographs, for the experience more than the artistic creation. Bears were rarely seen on the reservation and this one found me, rather than me finding the bear. I didn't see this one again, for if I had, I would have remembered him.

Seven years had passed since I had worked for the Lolo National Forest mapping elk habitat. That summer in western Montana's backcountry was chock-full of encounters with bears. One nerve-racking experience was deep in old-growth forest in the Sapphire Mountains south of Missoula. This fir and spruce jungle was so strewn with downed trees that I was often forced to travel along pathways of jack-strawed logs. Footing was always uncertain. When wet, the logs' jacket of moss and lichen proved as slippery as frog slime. Worse yet was the tendency of bark to peel from trees underfoot. Cautious progress and balance were indispensable assets for navigating these aerial causeways.

I began crossing a giant spruce nearly three feet in diameter. It spanned a steep-sided drainage incising the mountain. Arms outstretched like a tightrope walker, my eyes riveted on each step's placement, I inched along centering one step after the other and pressing through spiny-needled branches that poked from the trunk. Halfway across and some eight feet above a tangle of smaller logs below, I sensed something. Not a sound; not a smell. Just a presence. Raising my eyes from the scaly gray bark, I found a bear glaring back at me. Poised half erect, her front feet were on the far end of the same log. Worse yet, she had two cubs in tow.

I had been told that encounters with bears with small cubs were encounters of the worst kind. Given a chance, a black bear invariably will flee from a human—at least in places like this where they're hunted. But protective mothers may respond defensively at close quarters. A sow's greatest dread is confronting a boar black bear or grizzly that may kill her cubs. In short order she must choose to flee or stay put and fight.

My instincts said, *Be the baddest bear in the woods.* With no time to think this through, I went on the offensive. In my best scary bear imitation, I raised my arms and "wuffed" hoarsely at her several times—and more loudly than I realized I could.

She hesitated, glanced back at her cubs, and returned a menacing stare at me. *OK, this is it,* I thought. *Reduced to the bottom of the food chain.* To my immense relief, she turned and bolted into the forest, "wuffing" to her cubs who scurried up nearby trees. With my heart thumping like a sprinter's, I eased my butt down and straddled the log to regain composure. I now understood the meaning of the rock climber's expression, "sewing-machine legs."

My earliest experience with the black bear's more fearsome relative, the grizzly, was a didactic rather than menacing encounter. That same summer, my coworker Tad and I were mapping elk habitat on the flank of Big Hole Peak. A mile from camp one morning, we came across a meadow furrowed as if a mini earthmover had been at work. In mounds of freshly excavated soil, we found five-inch-wide paw marks. On the wider front paw, claw tips registered two to three inches beyond the pad—clearly not wimpy one-inch claws of a black bear. No, this was *Grizzly!* With their oversized claws, grizzlies unearth tender bulbs of spring beauties and biscuitroot. This bear, however, was past the salad course; he was digging for an entrée of three-ounce pocket gophers.

These small rodents tunnel networks of shallow subways, likewise in search of starchy tubers and roots. Raised ridges of soil, like a rodent road map, trace their passage across a meadow's surface. In underground larders the gophers stash harvested food that's crammed and transported in cheek pouches. These stores, too, provide a wholesome delight for the omnivorous grizzly when the more elusive rodents scurry away.

On a distant ridge, we spied a lone bear ambling toward the forest. The distinctive shoulder hump and shimmer of silver-tipped fur confirmed the first sighting of a grizzly for either of us outside Yellowstone and Glacier national parks. Pure wonderment tingled in my nerves at this near mythical sight. This was the animal that struck fear and awe in the hearts of Lewis and Clark's expedition members, for it recognized no superior. This giant's dwindling population was federally recognized as threatened with extinction in 1973. I could only marvel at the grizzly's perseverance under intense human persecution and commandeering of its domain. Was it possible this embodiment of wilderness might soon vanish from Montana, as had the wolf four decades earlier? Standing on its saucer-sized tracks I knew only that the bear was as much a part of this place as the spring beauties, gophers, and the mountain itself.

🦌 🦌 🦌 🦌 🦌

To many American Indians, bears are like humans. We are equally omnivorous and can walk on two legs. Both can be secretive or fierce, and both demand respect from other animals because of their power. I could relate. The Bull Lake bear did more than pulse adrenalin through my veins; he reinforced that I was on the right track. A study I had recently read reported that Idaho black bears killed 40 percent of newborn elk calves.[1] Odds were good that this was a biologically special area where female elk gathered each year to deliver the next generation. While I hunted elk to map secluded calving areas, this bear sought calving areas to hunt newborn elk.

With rising anticipation, I hoisted the tripod and camera onto my shoulder. I crept to a nearby cluster of firs—avoiding the path of the bear, but seeking a common quarry. As I slinked downslope on a bed of dew-drenched twigs and fir needles, the organic richness of decaying forest blended with the perfume of elk. Another ten minutes and I detected a familiar whisper of sound.

Elk cows and calves the author stalked in Bull Lake Canyon, June 1980

I concentrated, straining to sort the sound from the breeze slipping through tree boughs. A minute later, it came again—a faint "mew" from down the slope. Not like my cat Kintla's welcome-home greeting, but the plaintive call of a calf elk perhaps secreted beneath a snowberry bush by its mother while she made more milk from groundsels and sedges.

This was what it was all about—the serendipitous payoff of a wildlife biologist's fieldwork. As a GS-9 civil servant earning $20,000 and change a year, I had not been enticed by Uncle Sam with riches or fame. My constant reward lay in the next meadow, beyond the next mountain, in the life all around me. Mine was the prize of observing and recording Nature in places largely untrammeled by humankind.

Nearing the edge of the fir grove, I could see movement—russet and ginger slivers through the dim light of the last trees. Though they were mostly hidden by a bulge in the grassy meadow just beyond, I glimpsed heads, necks, and backs of at least eight elk. On my knees, then my belly, I wriggled to the forest edge, moving only when the animals dropped their heads to feed. As I crept closer, the rise ahead increasingly concealed their bodies. I paused to poke an index finger into my mouth and then raised it above me to test the air. On the side toward the elk it cooled more quickly and I reckoned, *I just might be able to belly-crawl to that shrub between me and the elk.*

Over several minutes, I painstakingly inched to a chunky snowberry bush. On the crest of the rise it shielded my approach from the elk, but likewise any view of them. *Were they still there?*

Of course they were; I could now hear their footsteps and teeth tugging grasses! And I knew that any noise or commotion would promptly send them running. So I abandoned the pack—and any chance for photo close-ups—nudged past the shrub, and in super slow-motion, raised my head. A stone's throw away, barely 10 yards, a cow elk towered above me with a newborn calf tucked between her hind legs. I was close, so close I could hear the calf's sucking while mom groomed its rump with long strokes of her pink tongue. In all there were ten cows, but no other calves visible. Shortly I spotted the ears of three others poking from geranium and bluebells on the far side of the herd.

At times like this, I felt like a skulking intruder, like a Westerner sneaking photographs of aboriginal people. So as cautiously as I had advanced, I retreated to the fir grove, snapped a couple of telephoto pictures, and slipped away with a final grateful glance. As I wandered down the ridge toward my camp beside the creek, a swarm of flies caught my attention. In matted grass lay the spotted hide and skeletal remains of a calf elk. Nearby lay two moist piles of hair-packed bear dung. This one's life had been cut short.

The following June, I returned to this hideaway blessed with lush milk-making vegetation, hiding cover for newborns, and security from humans. The elk came back also, validating my judgment of its nurturing import. As evidenced by fresh scat, Mr. Bear revisited too for the good grazing as well as the infant elk lottery. I returned just to witness life ongoing—renewal, survival, death to feed others, the continuum of life in the microcosm of a mountainside in Bull Lake Canyon. A decade later, I would study in detail elk reproduction and predation by black bears in another corner of Wyoming's elk country.

🦌 🦌 🦌 🦌 🦌

Elk always enthralled me, even on days when they seemed surprisingly ubiquitous. The backdrop of sunrise on snow-laden silvery peaks, though grand, never overshadowed the elk and other animals roaming the Wind River tribes' homeland. Certain images remain as clear as Wyoming's high altitude air. I recall one such sight minutes into a helicopter flight north of Dry Creek canyon.

Climbing a bulging ridge, we overtook nine bulls wheeling away from the rotors' approaching "thwack, thwack, thwack." All were big boys, sporting five or more tines on each antler. The white-tipped tines gleamed atop russet beams. The elk veered downslope, gathering speed. I watched waves of snow showered forward by pounding hooves, each wave an eruption of prismatic crystals glistening in the frigid air. As they loped single file, their black noses puffed nimbus clouds that passed backward from one muscled animal to the next. Twisting within my shoulder harness, I couldn't look away. Then our shadow slipped by and plunged with the ridge to Dry Creek canyon far below. Somewhere in its dark recesses, moose browsed willows in brisket-deep snow.

Notwithstanding their hard-to-miss size, moose were elusive even from the helicopter. This largest member of the deer family was also the rarest ungulate on reservation lands. I once saw 15 on a February flight—my personal best—while weaving up and down each of a dozen drainages spilling from the Wind Rivers. But that was exceptional. During three winters of surveys, I spotted just 1.5 moose per hour.

"How many do you count per hour?" I ask John Emmerick, the lanky, sandy-haired biologist from the Wyoming Game and Fish Department's Lander office.

"Depends upon the day, of course," he begins with the caveat that all biologists invoke to acknowledge daily variation in weather and animal behavior. "Last winter I averaged 22 moose per hour south of the reservation."

"I've hiked those canyons. Seems to me reservation habitat is similar," I observe.

"Yeah," John concurs. "Maybe even better, since the res' has fewer animals to browse the aspen and willow."

On another of my flights in late January, a trio of moose browsed firs dwarfed and twisted by blistering winds at 10,000 feet elevation. They appeared incongruous, like three Hershey's Kisses on a bed of whipped cream. Lost in a desolate world, they were safe from hunters. But what price did they pay for this snowbound security? Moose were figuratively trapped between the hunter's bullet and the icy grasp of winter.

*****

I found the moose the most intriguing large mammal on WRIR. A creature apparently designed by committee, it possesses gangly legs and a huge head adorned with a fleshy nose and a dangling throat dewlap, or "bell." Females wear no headgear, but males sport broadly palmate antlers.

Those of the biggest and oldest bulls resemble oversized serving platters fringed with steak knives.

In addition to its curious looks, the moose occupies a distinctive niche. It is the consummate browser, consuming 15 or more pounds of willow, aspen, chokecherry, and other shrubbery daily. Like all other hoofed animals that regurgitate and rechew their food, the moose is a ruminant with a four-part stomach. In the first two chambers—sort of fleshy fermentation vats—all food is predigested in rumen liquor. This is not Jack Daniels, but a soupy brew of bacteria and protozoans that lyse the cell walls of plants and free the precious nutrients within. By rechewing its food, the moose reduces particle size for the microbes that remix with the mash when it is swallowed once again. This multistage digestion scheme slows the rate of food passage, but its great efficiency serves the world's largest deer species well. With a shoulder height of nearly six feet and an even higher browsing reach, the moose exploits foods that its closest competitors only long to nibble. Aptly, its name comes from the Algonquian *moz*, loosely translated as "twig eater." On WRIR, moose favored willows; these constituted 92 percent of winter diets. Accordingly, I observed most moose in riparian habitats where willows hugged stream banks.[2]

Moose are largely solitary and unpredictable by nature. Surprising one can invoke flight, indifference, or aggressive behavior. The moose is the decider. A decade after my WRIR experience, I foot-raced a wide-antlered bull moose. Hunting elk one fall, I was hiking along a Forest Service trail in predawn light. Ahead, a glimmer caught my eye. A peek through my binoculars produced a moose with moonlight reflecting from polished antlers. The moose was headed my way on the trail; I was headed his.

The trail followed a thin crest of ridge. Dog-hair lodgepole swarmed the trail to my left. To the right, the ridge fell abruptly toward a tumbling creek. I decided to ease along that sage-studded ridge and pass him on the right. As I did, the moose dropped his head and charged. My hunting companion, who paused to watch, claimed it was one of the more entertaining episodes he had witnessed—me busting creek-bound through sagebrush, the moose closing in behind. As we lapsed from sight, Tim wondered if the day's hunt had ended before sunrise.

This was the peak of the rut and that hormone-boosted moose was apparently in no mood to share *his* trail. But after 40 or 50 yards, he broke off his assault, apparently satisfied with the outcome. Had he continued, he would have worn me on those wicked antlers.

"Nine on technical merit; three on style points," Tim awarded me. I gave him no points for hanging behind a tree.

Like this close encounter, the moose's most aggressive behavior is attributable to bulls during breeding season and fickle females tending young. But then, there are exceptions.

I was skiing one February day in Jackson Hole, Wyoming. My route would take me into deep snow county north of Shadow Mountain. Gliding serenely through an old Forest Service clear-cut, I spotted a cow moose crossing the logging road ahead. I stopped to watch her; the moose likewise detected me. She halted, I thought, just to look me over. At that point I had no reason for concern. It was mid-winter, there was no sign of a calf, and she was a football field away. But without provocation, she ambled straight toward me in a moose's deceptively speedy gait. I remained unconcerned, assuming she was just trying to scent me or gain a better view, but she halved the distance without slowing. This could be one of those statistical outliers, I reconsidered, that happen to those who spend too much time afield.

I planted my bamboo poles and sprinted hell-bent toward the edge of the clear-cut where the big trees grew.

Bull moose

The lodgepole pines regenerating in the cut block were spindly 8- to 12-footers, suitable for squirrels and pine martens, but not me. I fixed on a husky Douglas-fir that I knew I could climb, especially with this moose bearing down. Ms. Moose charged on, and I shifted to a higher gear.

I made it to the tree with time to spare, except that I was clamped to a pair of 205-centimeter-long boards. By the time I released the bindings, she was nearly on me. I grabbed the lowest branch and hurtled myself upward. She stomped one ski into the snow, studied me, and promptly left. *What was that all about?* I haven't trusted moose since.

🦌 🦌 🦌 🦌 🦌

As the Joint Business Council increased my wildlife survey funding, I substituted more helicopter flights for airplane charters. The reliability of the population data I assembled followed suit. The differences between fixed-wing and rotary-wing aircraft are many. Fixed-wing craft are champs for reduced ferry time due to their generally faster air speeds. Superior aerodynamics keep them aloft longer on a full load of fuel. They are cheaper to operate, maintain, and insure, and therefore to charter. And not unimportantly, single-engine airplanes come equipped with dual sets of flight controls. Even a novice could put one down should the chauffeur become incapacitated, I suppose.

Helicopters are easier on passengers in rough air. Their flexible top rotors smooth the bumps. Helicopters can fly slowly, even hover or back up, so animals can be observed more thoroughly for counting and classifying, even a second time if necessary to confirm a tally. In difficult situations, a skilled pilot can direct animals away from forest cover for easier counting, and even split a large group of, say, 200 elk into smaller bunches. To limit stress on animals, a pilot and crew must accomplish this with care. It's true that animals are quicker to flee from an oncoming helicopter, as if a giant raptor is about to have them for lunch. But given their maneuverability, helicopters often spend less time positioning biologists to classify a group of deer than an airplane, which may need to circle several times.

The small ships in which I flew provided exceptional visibility—like looking outward from a goldfish bowl. Bulging plexiglas wrapped the top, front, and sides of the cockpit, even offering a partial view below. To see passing animals from airplane windows that began at almost shoulder height, I sometimes found my forehead pressed against the glass.

Helicopters safely fly low to the ground. This makes them useful not only for doing wildlife surveys, but renders them indispensable for capturing animals with net guns or immobilizing drugs propelled from dart guns. Try *that* from an airplane! Landing nearby to tag the beast is the crucial next step. The walk will be shorter after landing a chopper than a Cessna. And speaking of landings, for us coffee drinkers or anyone with a wee bladder, impromptu pee stops in a helicopter are a cinch.

🦌 🦌 🦌 🦌 🦌

Ron Gipe owned and flew a helicopter out of Polson, Montana. I contracted him to fly winter surveys of big game over the Wind River Range and Owl Creek Mountains during the winter of 1978–79. He trailered his Bell 47 to the Red Rock Lodge, a covey of log cabins pinched between the rosy sandstones of Red Creek and the Wind River. It offered a central location from which to survey the Winds and the Owl Creeks. It was also decent lodging and meals for a pilot, and for me when I chose to forgo the hour-long drive from Lander at oh-dark-thirty a.m.

I have flown with many first-rate chopper pilots. Ron was one of the best, though occasionally keen to demonstrate talents I didn't need to experience first hand.

Ron described a helicopter as "ten thousand parts loosely moving in unison." That succinctly accounted for the exacting maintenance schedules and high insurance costs. It also crystallized the need for a pilot's attention to any odd odors, shudders, or "whirs" that would possibly require an unscheduled landing. Furthermore, those jet- and piston-powered engines have been known to halt without notice. Or as economist Herbert Stein observed, "If a thing cannot go on forever, it will stop."

We had finished the first leg of a morning big game survey over the Owl Creeks, and were headed toward the airport in Thermopolis, Wyoming, to refuel. The conversation had drifted from the scarcity of mule deer and elk in such agreeable habitat to the peculiarities of helicopter flight.

"Did you learn to fly in the military?" I asked, since that's how many chopper pilots learned their trade.

"Yeah, Army," he replied nonchalantly.

Judging that he was about my age, I asked, "Were you in 'Nam?"

"1969 and '70," he said.

"The same years I was there," I replied, feeling that instant connection you get having shared a similar experience with someone, even if not together.

"Did you fly Hueys, or Cobras, or something bigger?" My curiosity rose, now envisioning the details of each drab green machine. As though reliving a former life, I recalled the differences in rotor cadence, and felt my hair bristle from revived memories flashing before my mind's eye.

"Hueys, gunships, and slicks," he said. "Were you Army?"

"Marines," I replied, probably with a tinge of pride in my voice. "'I' Corps, from Da Nang into Laos."

"Infantry then?" He glanced at me as he swung the chopper left and we watched three fleeting mulie bucks. They bounded across the amber grasslands of Kate's Basin toward a nearby ravine and dove into a thicket.

"Yeah, machine gunner. But I spent some hours in 46s and 53s being hauled from one shoot-'em-up to the next," I said. "When I was medevaced, a Huey took me out."

In vivid detail I recalled Huey and CH-46 Sea Knight choppers coming under ground fire, especially during emergency medevac extractions of wounded Marines. "You must have seen a few hot LZs while you were there," I said with a question in my voice.

"A few," Ron distantly replied.

"I can imagine," I mused.

"You ever on board when a chopper was hit?"

"Once. One time on a Sea Knight bullets came through the floor." I visualized Marines packed nine to a bench seat along each side of the ship, facing each other across a five-foot divide. "Two rounds burst through, spraying metal inside the chopper. The guy next to me took shrapnel in his leg."

"Close call," Ron murmured from a place in his past far away.

"Yeah, close call. But we made it out OK," I whispered, tasting the panic of the moment.

Fate, luck, or a higher power had been on my side. Fortunately, I had never been on board when it happened, yet I knew from firsthand accounts that engines, transmissions, and others among those 10,000 parts were sometimes knocked out by enemy fire. In that case the pilot "autorotated" the ship to the ground—a kind of controlled, gliding crash. Even without power, the rotors continued to rotate, braking the descent and reserving the pilot some measure of navigational control. Oftentimes, the crew survived the landing, assuming the machine didn't burst into flames on impact. Other times it didn't work out that way.

"It seems like such an advantage," I said, "being able to land more or less vertically in a pinch." The thought made me feel more comfortable flying wildlife surveys in helicopters than airplanes. In an emergency, airplanes can glide quite far, depending upon present altitude. But they need a long stretch of unobstructed landscape to land, with power or without. They can't just drop from the sky onto a postage stamp forest clearing or streamside gravel bar.

When my tribute to helicopter flight ended, Ron affirmed, "That's one of the things I like about flying these outfits."

Then he glanced at me. "Have you ever done an auto-rotated landing?"

Before I completed "No," he cut the engine power. I felt my windpipe shorten and watched the metal clipboard levitate off my lap. He took it down to 200 feet above ground level, I suppose, and throttled the engine again. The descent slowed perceptibly and the Bell 47 fluttered to a rest on the airport tarmac.

A grin creased Ron's face. "Now you have."

I just grinned back.

5

Stranded

Long cobalt silhouettes linked the sparse junipers that slipped beneath as we chased the helicopter's shadow across dissected sagelands. An immature golden eagle wearing white-banded tail feathers, the decorative plumes prized by Plains Indians, streaked past the left door where I was seated. The dense, still air made for ideal flying conditions. It was a great day to be alive, soaring with the eagle.

Our pilot John guided the Hiller 12E around the east flank of Black Mountain, so named, I assumed, for its cloak of lodgepole and limber pines that shaded to pyramidal firs and spruce above 9,000 feet. This 10,177-foot hulk dominated the skyline. Behind Black Mountain lay Crow Creek basin, where a pretty willow-lined stream nestled between the 11,000- to 12,000-foot-high crests of Black Ridge on the west and Trail Ridge on the east. These ridges joined at the north, forming an elongated horseshoe that fed Crow Creek's waters over 2,000 feet below.

Our mission on this subzero morning in January 1980 was to survey elk, mule deer, and bighorn sheep in the Owl Creek Mountains. Helping me was Rawlin Friday. Rawley was Arapaho and a tribal game warden. About my height but stockier, he could handle himself. I liked flying with Rawley. He was devoted to the reservation's wildlife, made a

The author on Trail Ridge glassing for bighorn sheep, June 1979

jovial companion on surveys, and owned an iron stomach, something others I had flown with didn't possess.

In 1944, helicopter pioneer Stanley Hiller built his initial rotary-wing aircraft at age 18. Hiller Helicopters' first production aircraft, the Hiller UH12, first flew in 1948, the year I was born. More than 2,300 were built for commercial and military use by 1965. The next rerun of *M.A.S.H.* you watch, look closely at the choppers used for medevac or for transporting Hawkeye from the 4077 in the series finale. They are UH12s, which the military first purchased in 1950 as H-23 Ravens.

Increasing altitude hampers performance, as with all reciprocating engine craft. Our Hiller was equipped with a Soloy turbine conversion to remedy that limitation. The one drawback was the turbine's increased thirst for fuel, giving us only two to two and a half hours aloft per fill-up. We carefully planned the day's work with that in mind.

It was now mid-morning. Nearly two hours had elapsed since Rawley and I had met at the Thermopolis airport and begun the day's survey. We had already recorded 129 mule deer and almost 200 elk across the eastern three-fourths of the sixty-mile-long Owl Creek Mountains—improved numbers for that area compared to last winter's counts. The high country loomed ahead.

I had planned the day's flight to conclude with the mountainous area surrounding Crow Creek. We would be operating at our loftiest elevations when we searched for high altitude elk and bighorns along Trail and Black ridges. Unlike sheep in the Wind Rivers, which migrated to lower-elevation cliffs in winter, bighorns here wintered along wind-scoured ridgetops and escarpments. We carried another 40 gallons of fuel in five-gallon jerry cans in the Hiller's twin cargo baskets—one mounted on the skid beside each door. The additional weight would reduce performance at high altitudes, acting like ballast on a submarine, but would avert a gas-guzzling ferry to refuel in Thermopolis. This dance of performance versus mission time vexes all remote mountain flights. But as one pilot told me, noting that far too many aviation accidents are caused by running out of fuel, "The only time there's too much fuel on a helicopter is when it's on fire."

After completing our counts of elk and deer at the lower elevations, John throttled the Hiller toward Trail Ridge. The fuel gauge now registered one-quarter full. We would land on the ridgetop high above, refuel with the jerry cans, and begin our hunt for bighorns.

As we approached Trail Ridge, John unexpectedly settled the Hiller in a foot of snow on the Crow Creek road. His voice was tinged with concern as he said, "I want to check on something." Noting our probing expressions, he added, "I smelled something."

As I noted earlier, weird sounds, vibrations, and smells can be telltale signs of trouble to a helicopter pilot, like the tingling sensation I had felt in the Wind Rivers before lightning nearly struck me. I hadn't flown with John before, but his attention to this anomaly eased my usual concern about new pilots. Without shutting the engine down, he climbed out through the flimsy acrylic door and examined the Hiller's mechanics. In minutes he was buckling into his shoulder harness, informing his onlooking passengers, "False alarm. Everything's OK."

Rawley and I shared relieved looks. "Good," I said as the engine whined and snow whipped like meringue around the ship. "The closest service station's a healthy hike from here."

As the Hiller whooshed forward, Rawley and I resumed scanning for big critters or the telltale tracks of their recent passage. I shuffled the folded stack of topographic maps on my lap, bringing the Monument Peak topo to the top. It was mid-winter. A storm had dumped fresh snow—conditions that covered old animal tracks and maximized visibility of game. The Hiller strained and shuddered upward into a world above 10,000 feet that was snowbound except for stunted treetops, cliffs, and shreds of windblown ridge. This was the windiest area of the reservation. Funneling eastward off the Continental Divide through the Wind River valley, winds howled unobstructed before encountering Black Ridge. On windward slopes and ridgetops, velocities could be brutal. On the leeward flanks, turbulence was particularly dangerous, sometimes insane. In sum, advancing weather fronts made surveying game impossible. Today, high pressure had settled over Wyoming bringing favorable conditions. Yet John was busy at the foot pedals adjusting for the swirling gusts that yanked the tail boom left, then right. Through the noise-dampening padding of my flight helmet, the main rotor pounded like a jackhammer.

We intended to refuel along the crest of Trail Ridge, which offered level landing sites. The upsurge in wind now threatened that design. A misstep along the ridge's narrow spine could make for an untidy landing. A rogue downdraft or wind shear could prove disastrous. John nudged the Hiller up the west side of Trail Ridge. The ship sailed skyward on updrafts, then yawed sideways as the air settled beneath the rotary wing and buffeted the fuselage. Gaining the ridgetop, the Hiller suddenly banked west across Crow Creek Basin. "Thwack, thwack, thwack!"

Sensing our intense stares, John announced, "We better look for a landing site somewhere below the ridgetop. Somewhere the winds are more manageable than up here."

The other option, descending over 2,000 feet to Crow Creek to refuel, would cost us precious fuel to regain our 11,000-foot altitude. Neither John nor I wanted to do that. We circled back and scanned the bleak landscape below the ridge crest.

"Maybe there!" John jutted his chin forward.

Rawley and I peered in that direction, then back at John's eyes for confirmation. "That spot?" Rawley questioned.

Just above the altitudinal limits of the subalpine fir belt, our eyes converged on the only level place—more like a possibility of a place to land. A spur off Trail Ridge descended some 200 feet to a flat spot, then plunged to Crow Creek far below. The flat was snowbound like everything else, but offered a landing perch large and stable enough for the petite helicopter to roost—or so we reassured ourselves. Fresh avalanches shredded Trail Ridge's winter mantle to our left and right. Tons of snow had plunged hundreds of feet in some slides. A spasm of unease clutched my chest.

John circled twice more to gauge the winds. "You OK with this?" he purposefully asked, not taking his eyes off the mountain and the fast-approaching treetops.

A glance showed the fuel gauge had dipped below one quarter. From just 100 feet above, the little landing zone looked safe enough to me.

"Yeah," I replied. "Let's do it."

A brief lull in the wind aided our approach. Into the snow the Hiller settled; the cargo baskets buoyed it from sinking fuselage-deep. As the turbine engine wound down, the three of us unbuckled our harnesses, stepped into the cargo baskets, and then into thigh-deep powder. Bucking the snow, Rawley and I ferried the jerry cans to John, who emptied the contents of each through its flexible pour spout into the fuel tank. Another two hours of flight time awaited—enough to thoroughly survey the 20 miles of lofty horseshoe surrounding Crow Creek, as well as Mountain Meadows below, and return to Thermopolis with a helping tailwind.

Back in the Hiller, we rebuckled harnesses, squeezed into helmets, and connected dangling avionics plugs. John summoned on the intercom, "Ready?"

Rawley and I each replied, "Ready."

"Clear!" John announced and fired the turbine, which after several tries rumbled to life. Sluggishly at first, and then with increasing authority, the Hiller rose from our perch, as great eddies of snow swirled around us.

When a helicopter lifts off, it begins in HIGE (chopperese for "hover in ground effect"). This signifies a condition of optimal performance encountered when operating near the ground. A helicopter requires less power to achieve the same amount of lift near the ground because the ground interrupts the airflow beneath the helicopter, reducing the downward velocity of airflow and translating to more vertical lift. Essentially, a compacted cushion of air between the rotary wing and the ground supports and lifts the chopper. Once a height exceeding one rotor diameter is reached, the helicopter transitions to HOGE, or "hover out of ground effect." A pilot will then increase the angle of attack, or the rotor blades' forward tilt, to develop forward moving speed. Continuous, coordinated adjustments to the throttle, foot pedals, and cyclical and collective controls keep the contraption from gyrating aimlessly.

I mention this not to impress you with my elementary grasp of aviation physics, but to illustrate a point. Flying a helicopter is a complex affair, demanding a pilot's constant attention. Airplanes are designed to land on runways, or, in the case of floatplanes, on water. Doing otherwise isn't impossible. Airplanes can land on roads, in fields, and less suitable locations in a pinch. Alaskan bush pilots land most anywhere they can. Larry Hastings once landed me on a faint two-track road near Crowheart Butte in his Cessna 182. We both had to pee so badly, after too much coffee and nearly four hours of counting pronghorns, that bladder damage seemed imminent.

Helicopters are designed to land at unimproved landing sites. Think of some butt-puckering situation—a pocket of grass pressed between towering firs, an old burn strewn with snags and downed trees, or a sage-covered hillside demanding a one-skid touch-and-go in a gale—and some pilot has landed there. Such landings I had survived in Vietnam I considered heroic. In natural resource work (firefighting and rescue work in particular), they happen with equal regularity. It's just that in civilian employment, rarely are folks shooting at you.

Ours was not the classic takeoff I described three paragraphs earlier. John executed a variation, God bless him, maybe because we were perched on this little launch pad so near an abruptly rising slope; maybe because the swirling snow hampered visibility; maybe because he felt the breeze nudge the helicopter and wanted to clear all danger and then turn into the wind; or maybe because of some technical aspect that I probably couldn't explain if I knew it. I don't know the reason—never thought to ask him.

Instead, John took the Hiller skyward without dropping the nose to gather forward speed. We went straight up. Up, briefly—until we fell from the sky. Not like a rock, rather like the autorotation that Ron Gipe had demonstrated a year earlier. In fact, that is what the Hiller did; it autorotated right back to our launch pad with a jolt and a billow of snow.

It happened so quickly that I didn't have time to get terrified, although every time I have thought about it since, my heart zips a bit. Ten yards to one side or the other, and we would have alighted beyond that flat spot on the spur ridge—and then begun sliding and tumbling creek-bound. The dense stand of firs just below would have halted our descent, but the damage caused by the rotor blades battering the mountainside would be done by then. Picture locking an industrial Mixmaster on high speed, releasing the handle, and then watching it dance from a countertop onto the floor—only with way more drama. Helicopter rotor blades are huge. Each measures almost 18 feet in length on the Hiller. And at the full-throttle revolution needed to lift the ship and passengers from the ground, they wield tremendous force.

We might have survived, if the cabin remained intact. I don't know the fracture resistance of the acrylic canopy or the tensile strength of the supporting metal alloy structure, but both would surely have been tested that day. Fire is the other life-threatening danger, particularly for unconscious or disabled passengers unable to release safety harnesses.

None of these "could haves" happened. We plopped right back where we had been moments before. Rawley and I looked wide-eyed at each other and then at John.

"What happened?" I started.

"Lost the fuel pump," John replied. "Just conked out!"

"The fuel pump!" Rawley exclaimed. Then he paused, and inquired, "Can ya fix it?"

"I don't know. Probably not here," John responded.

I looked at my watch. It was 11:30—plenty of time to check out the fuel pump and resume our work. Of course it could be fixed, I reckoned. Besides, we were snowbound at 10,000 feet, and miles from anywhere.

Everyone unbuckled and got out as the rotor blades spun down. "Whoop, whooop, whooooop." Their tips assumed a sad, lifeless droop. John rummaged in the tail boom's stowage compartment and came out with a small tool box. He climbed up to the engine and fiddled with something for a few minutes while Rawley and I watched and asked a couple of pointless questions. When John stepped down into the cargo basket, he looked grimly past us.

"Let's get back inside," he ordered.

After he triggered the ignition, the engine sputtered but failed to fire. Again he tried to no avail, then blurted, "No way. Can't be fixed here. Most likely, it will need replacing."

I felt the air go out of my lungs. My first concern was that we wouldn't complete today's survey. The second was that we probably wouldn't be finishing up tomorrow either. And third, I wondered how long before we would be rescued. That's the thought that stuck.

Before we climbed back into the cabin, I got my daypack out of the stowage compartment. I had water, a sandwich, gorp, and a chocolate bar in there, along with gaiters and a parka strapped to the outside. Flying in the heated helicopter, I was too warm to wear the down parka. It sure was welcome now.

Inside, Rawley and I removed our flight helmets. I replaced mine with a stocking hat. Rawley donned a blue wool cap with earflaps. John fingered the radio switch, adjusted his helmet's mike, and tried to raise the Riverton or Lander airports. "Lander unicom, this is helicopter ..."

He repeated the transmissions a dozen times. No response. He glanced at us, probably wondering what we were wondering. Then he retuned the radio to another frequency, the Mayday channel. The vapor trail of a jetliner was passing overhead. His determined efforts to raise a reply were again fruitless.

He wrenched the helmet from his head and offered a couple of choice expressions—the insightful kind appropriate for our now apparent predicament. He turned to us and said, "We have no radio contact. The avionics must have been damaged when we landed."

The radio antenna was buried in the snow compacted beneath the fuselage. Even if we dug it out, John replied to my inquiry, it would make no difference. I recall disliking his defeated tone. Rawley and I were willing to dig, excavate, burrow—whatever was needed. But John's further explanation of the problem convinced us that digging was just a good way to get soaked and then good and cold—a surefire formula for hypothermia.

I recall the sun beaming through the clear canopy. It felt warm on my face at its midday zenith. But in five hours, it would sink behind the Wind Rivers along with the air temperature. The morning forecast on the Lander AM station called for 20 degrees below zero tonight. That was at 5,400 feet elevation, not here at 10,000 feet. The helicopter's round thermometer showed just 12 degrees on the plus side. Without an insulating cloud layer, tonight's temperature would be bitter.

John tried the radio again as I mulled over our options. I knew that convention dictated remaining with the aircraft in the event of an accident. Search craft would look for the downed Hiller, guided by the onboard ELT, or emergency locator transmitter. Every airplane and helicopter is equipped with one. A crash landing's impact activates the ELT, which emits a radio pulse signal on the 121.5 megahertz frequency. ELTs can be manually activated also. John did so, recognizing that our moderate impact may not have done the job. Any aircraft monitoring the ELT frequency and flying near enough to receive it would be alerted to our emergency. Great! We would be rescued—as soon as somebody knew we needed rescuing.

These were the days before the federal Office of Aviation Services required flight following of government-chartered craft—15-minute-interval communications between small aircraft and radio base stations. Before those requirements were implemented in the 1990s, dozens of pilots and passengers in downed airplanes and helicopters might have been saved had their accidents not been detected belatedly. Over my 30-year career, seven of my colleagues would lose their lives while conducting wildlife surveys or radiotelemetry missions. The small plane carrying one of them disappeared over the Arctic Ocean during a polar bear census. Another went down in the mountains of Arizona while surveying desert bighorn sheep. Two fixed-wing pilots, with whom I would log many hours tracking radioed elk, died on ill-fated missions just days after I had flown with each. Still others died in helicopter crashes. And another wildlife student I accompanied on my first wildlife flight back in graduate school—the flight that introduced me to the fundamentals of aerial radiotelemetry—died a few weeks later. The Super Cub in which he was radio-tracking elk augered into a dense stand of Montana lodgepole pine. It wasn't found for two weeks. All of these and many more flights that took the lives of biologists and wardens and pilots I never knew testified to the danger of this aspect of wildlife work.

In most downed aircraft incidents, time is of the essence. The only exception, of course, is when there are no survivors. Injured passengers may suffer shock, blood loss, life-threatening trauma, and the risk of death due to exposure. The clock starts ticking immediately. The first 24 hours are critical. The sooner help arrives and the injured can receive medical treatment, the greater their chances of survival. Consequently, rapid detection of a crash and its location are paramount.

We were *so* fortunate. The three of us suffered no injuries. But I knew that a night spent at tonight's temperatures—likely 30 below zero—spelled hypothermia. I also knew the following: Rawley had mentioned to the aviation tech that fueled the Hiller in Thermopolis that we were doing a wildlife survey toward the East Fork of the Wind River. But he likely wouldn't be alarmed if we didn't return for fuel. It wasn't his responsibility. My office wouldn't be alarmed until nightfall, because I had planned other fieldwork after the flight. John had filed a general flight plan: "five-hour wildlife survey in the Owl Creek Mountains." He filed the plan with his employer, Hawkins and Powers Aviation, before leaving Greybull, Wyoming, an hour distant from here by airplane. When would H & P consider John overdue in Greybull? Finally, our exact location in this sixty-mile-long mountain range would take time to pinpoint should our ELT fail. I learned later that ELT failure rate was disconcertingly high.

But most disturbing, I learned that the standard winter emergency equipment—food, water, sleeping bags, and snowshoes—was missing from the stowage compartment. Besides the tool box John retrieved, the compartment contained a flare gun, an oil-soaked blanket, and nothing else. *Why?* A serious oversight on someone's part. With three sleeping bags to crawl inside, or even a couple to drape around us, the night would have been bearable—certainly survivable. Considering these grim circumstances, I was contemplating another plan.

After a half hour of listening to unanswered radio transmissions and sharing my gorp and water with the others, I was ready for action. I could already feel my toes chilling with inactivity in the unheated cabin. I attributed my dreadfully poor foot circulation to countless days in too-tight hockey skates on Michigan lakes and outdoor rinks. With the sun arcing westward and 20 more hours before sunrise tomorrow, time was wasting.

Likewise, Rawley was not keen on death by freezing. We shared the same thoughts. "I think we ought to hike outta here," I stated.

Rawley's nod brought a strident protest from John. "We should all stay together, all stay with the helicopter. That's where they'll look for us."

I agreed with the first part. Sticking together was imperative, whether we stayed or left. I just believed we should all hike out, helping each other as needed. Rawley and I knew from the maps I carried that the nearest help was the Duncan place, a ranch on the East Fork of the Wind River along the reservation's western boundary. We also knew that it was 2,500 feet below us and 12 miles west. A reasonable task in summer, I thought, but certainly an ordeal in untracked snow, and without snowshoes. Furthermore, part of the slog would happen after dark, even if we set out immediately.

But on the plus side, after dropping 2,000 feet down Trail Ridge to Crow Creek, there was a two-track to follow the entire way to the East Fork. Sure, it would be snowbound, but an obvious course, nonetheless. Rawley had traveled those tracks since childhood. I had driven them a couple of times in summer. We also knew that just two miles south of where we would intercept the two-track paralleling Crow Creek, a cowboy cabin was nestled beside the creek. Rawley recalled a potbellied stove inside. Make it there, and we'd be comfy.

The debate wore on. John insisted that a search craft would soon be wheeling overhead. My mind was racing with more pessimistic scenarios. I argued that if the search craft were a fixed-wing rather than a helicopter, we would have to wait an hour more for a helicopter. Then there was no assurance a helicopter could land up here after dark. We could remain stranded till daylight. And what if a search effort wasn't dispatched today, or didn't find us for any number of reasons? Even without the service of snowshoes, I knew I was fit enough to trek out.

John shivered. His leather aviator jacket was lighter than our parkas. I took that as a sign that this forty-something former LAPD pilot dreaded the prospect of enduring a night at 30 below as much as me. He was weakening to option B.

Rawley and I shared the logistics of the undertaking. "We'll follow this spur ridge down to Crow Creek." I traced the route on the map with my finger.

"Jim Hill's cabin is about here," Rawley pointed. "The mice won't mind some company."

"The Duncan place is over here. It's the first place where we'll find help," I added as John's eyes widened at its considerable distance from the cabin.

"I don't know," John hesitated, clearly conflicted. "You sure there's a cabin there?" he questioned Rawley. "I don't see it on the map, like the Duncan place."

"Yeah, right about here," Rawley pointed, then eyeballed him. "They don't put every little shack on these maps."

"John, this is doable," I assured him.

Despite the effort of slogging through bottomless powder, gravity would be our ally. It was all downhill to Crow Creek. Before leaving the helicopter, I studied the map and penciled the least grueling route down Trail Ridge. Minutes later, we all plunged into the thigh-deep snow, secured the helicopter rotor with the tie strap, and set out.

John and Rawley hiking down Trail Ridge in January 1980 (the disabled helicopter is visible on the skyline)

I broke trail; Rawley and John followed in the furrow I plowed through the snow. I couldn't decide if it was harder to raise each foot nearly out of the snow before plunging it forward again, or to just push the snow forward with my legs as I gained another two feet per stride. So I executed a few minutes of one routine, then the other. No score for style points here.

Within minutes, we were all breathing hard. We had covered barely 200 yards—*maybe* 20 minutes of toiling—when Rawley shouted, "Hold up!"

I don't think Rawley would mind me volunteering that he was on the heavy side. He enjoyed fry bread and other Arapaho fare not generally considered health food. But he was an outdoorsman and was younger than John and me. I expected him to keep pace or maybe trade off the lead, so I was surprised when he called for me to stop. I turned to see him a few yards back, supporting John who was bent and clutching his leg.

"You doing OK?" I called.

"Damn cramp in my leg," John grimaced.

I trudged back and located the outline of a downed tree mounded with snow. I brushed it off and Rawley helped John to it. Sitting hunched on the log, John began rubbing his aching left hamstring. After a few minutes, he said he felt better, and we continued on.

Another ten minutes and John cried out. He had cramped again. Rawley and I helped him to another log that we cleared. This time John was really hurting. Both hamstrings were agonizingly cramping, and he was gasping for air. He began massaging one leg and I offered to help with the other. I unzipped my parka and retrieved the quart water bottle from an inside cargo pocket. I had filled the half-empty bottle with snow before we left the helicopter. Eating snow can ward off dehydration. Unfortunately, the energy required to melt it can sap strength, reduce core temperature, and hasten hypothermia. I was hoping to thwart those effects by warming the snowmelt before drinking. Next to my body, the snow melted faster than in my daypack. Guessing that John was dehydrated, I encouraged him to drink freely of the bottle's slushy contents.

The muscle bundles in John's legs were bound in knots. Kneading gradually relaxed the tightness in his hamstrings. John thanked us profusely for our forbearance, apologized for holding us up, and said he was ready to continue. This scenario repeated itself periodically. When he was moving, we made decent progress. He just wasn't moving consistently, repeatedly hesitating to catch his breath and stroke his thighs. Standing six feet three inches tall, John sported legs rivaling a *Star Wars* Imperial Walker. With legs that long, he should have handled snow like a bull moose. Instead, they were a bane, afflicted with excruciating cramps. I empathized with John, but was increasingly anxious about the hour. He was obviously not rehydrating, or his physical conditioning was abysmal. Either way, our progress was distressingly slow.

Somewhere down the mountain, Rawley's eyes and mine locked. Neither of us spoke, yet we read each others' transparent thoughts. *If I'd had any idea that he would be this slow ...* Sooner than expected, nightfall would engulf us. Nonetheless, there was no wavering, no question of resolve in either of our faces. We were both committed to going on, especially now that the helicopter was hundreds of feet above us. Returning to the Hiller was *not* an option.

So we went, ever downward, as the sun receded southwest toward the dazzling summits of the Wind Rivers. Eventually the dimly lit pine and fir gave way to russet and golden willow thickets demarcating the course of Crow Creek.

"Doin' OK, John?" Rawley called as he and I waited at the forest margin. Miles from the nearest combustion engine or humming power line, only John's labored breathing and the swooshing of his feet through the snow breached the stillness. We *had* to reassess our situation.

"We've got two choices," I rationalized. "We can either continue on together and hope to get Hop-along to the cowboy cabin before we're all spent ..."

"Or," Rawley interrupted, "before he has a heart attack."

"Or we can split up," I whispered, as John's panting grew louder. "One of us stays with John, the other hoofs it to the Duncan place for help." The ranch's snow machines would get assistance into Crow Creek in no time. If the worst happened, we could bundle John in a sled and tow him behind one of the machines. Nine miles, the distance I estimated from the maps, was nothing by Arctic Cat.

Rawley and I looked at each other a long moment, gazed toward the rosy summits of the Winds, then back toward John's raspy approach. Neither of us wanted to make the call.

Finally, Rawley spoke. "I'll stay with John. You go ahead."

"You sure?" I asked.

"Yeah. You're in better shape than me. You'll get there faster," he grinned.

I sipped the remaining liquid, then thrust the plastic bottle and squashed sandwich I had retrieved from my daypack toward him. "You'll need this more than me. You can fill the bottle in Crow Creek." Rawley refused, then with some prodding acquiesced. He took the bottle, casting a glance at John's pathetic figure emerging cheerlessly from the forest. "You keep the sandwich," he resolutely replied, refusing to take it.

We explained our intentions to John, calmly detailing our logic. He objected only mildly, beleaguered by his part in our predicament. One last encouragement to both, and I slipped into the willows.

Along Crow Creek, the snow was barely knee deep and still fluffy. I wove through the tangled stems lining the frozen stream. At the creek, I stopped for a long pull where water bubbled through a gap between ice-covered rocks. It was mouth numbing, but I drank as much as I could until my scalp began to ache. I found the road along the west side, where a porcupine had left the telltale drag of its tail. I veered south. The remaining 11 miles from here to the East Fork were gently downhill. I would make good time if my energy held out.

I reflexively swung one pac boot and then the other through the scattering snow. The adrenaline was pumping nicely. I forced myself to concentrate on what I needed to do. Still, I couldn't avoid stewing about how the others would fare, whether splitting up was the right decision, and whether I would find someone home at the Duncan place. *How far was the next closest ranch?* Those thoughts conspired to distract me from the long trek ahead.

In less than an hour, I covered the two miles to the road junction where I turned west and left Crow Creek behind. I couldn't see the cabin, but I knew it lay just a few hundred yards farther downstream. I considered going there myself, getting a fire burning in the stove, then heading back to help Rawley. I thought better of it. That is not what we decided, and for good reason. If John's cramping and exhaustion worsened, he would be better served the sooner I got him help.

🦌 🦌 🦌 🦌 🦌

It was shortly after 5:00 p.m. The snow now matched the height of my 10-inch boots. I paused to take in the last light glowing crimson atop Trail Ridge behind me. Searching, I couldn't spot the Hiller, apparently concealed by fir trees or their inky shadows. I mouthed several scoops of snow, allowing my tongue to rewarm after each. Suddenly, I began to salivate, remembering the Milky Way tucked in my pack. With the sandwich long since devoured, I had hoarded this final item for a later time when those 250 calories might prove vital. Now seemed as vital a time as any.

Just ahead I noticed a single set of elk tracks crossing the roadway. As the last of the chocolate and caramel dissolved in my mouth, I saw where the elk had pawed and cropped Indian ricegrass, identifiable from its airy seed stalks. Over there a fringed sage was nibbled, leaving the animal's nose print pressed in the snow. The tracks meandered and vanished into the indigo snow at last light. Past the shadows and stillness, my thoughts reenacted the timeless pounding of hooves.

Following the somber silhouettes of sage and rabbitbrush down the snowmelt-scoured terraces to the Big

Wind, then continuing along its cottonwood gallery forest through badlands and prairies, I envisioned the ancient migrations of buffalo across the centuries. Gracing the breadth of the Great Plains, the great herds thundered and pulsed in nomadic rhythms in concert with winter storms and summer rains. Behind them on foot and then mounted on painted ponies, the hunters likewise shadowed the seasons and processions of *boha-guchu* (Shoshone for "bison").

Over the intervening century, the migration trails had faded or were trampled to dust by their beef-bearing cousins. Now, restoring these largest of native herbivores seemed as unlikely as the return of woolly mammoths. The stockman's disdain for competition and the bison's penchant to roam to the horizons—without regard for rivers or roads, fences or farms—had sealed their modern-day fate. Only in Yellowstone National Park, Jackson Hole, and Utah's Henry Mountains were remnant herds given license to live unconfined. Unlike the Wind River's resident elk, deer, and pronghorns—temporarily dimmed like sunlight by a transient cloud—the buffalo economy had vanished forever.

Why did *I* mourn that world's eclipse? They were not my ancestors, neither the buffalo nor their pursuers. Being raised on the shores of Muskegon Lake (the Ottawa Indian term *masquigon* means "marshy river") provided no nurturing connection to native people I had never met. *Maybe*, I mused, *I was hardwired with compassion for all fellow creatures*. Not likely! I could be as insensitive as anyone else. Instead, I concluded that my feelings arose from a common purpose. My work and time among the Shoshone and Arapaho people fostered this bond with their present and veneration for their past. Glancing toward Black Mountain's towering hulk, I felt deeply thankful for this opportunity I had been granted.

Squinting at the cloven imprints left not long ago, I considered how comfortable the elk was in this environment.

Her night vision, superb hearing, and sense of smell I could only envy. Insulated by subcutaneous fat, a thick hide, dense underfur, and a weather-repellent outer coat of hollow guard hairs, the elk was well suited to a life among the elements. I, on the other hand, felt like a vulnerable speck beneath the eternal night sky on this lonely frozen land.

With no concern for my welfare, I had left Rawley and John in the gathering dusk. Now for the first time, thoughts skittered through my head that were material for fictional thrillers. Somewhere in the night, mountain lions roamed the canyons and forests seeking prey my size and larger. *But you've traveled alone in lion country many times before,* I chided. My graduate studies of mountain goats found me mountaineering and camping for days at a time miles from anyone among the peaks and snowfields atop the Selway-Bitterroot Wilderness Area. All that differed now was my itinerant imagination. Besides, things could be worse. The black and grizzly bears that frequented this region were sequestered in their winter dens, fast asleep.

While I was moving, the swoosh of each step drowned the silence—that place where creatures of the night abide. Still as a fence post now, I remained a captive of the ethereal peacefulness a moment longer. I listened to the night and thought I could as well be among the hovering stars.

I stirred with a shiver. It was cold, very cold. Well below zero, I ventured. The wind had long since failed, for which I was grateful. Exposed as my route was, a bitter wind chill would have mercilessly sapped me. Still, my breath caked my beard and mustache with icicles. *Time to get hustling,* the voice within implored. To some internal cadence, following one track of the road, then the other, I methodically plodded along.

It must have been seven o'clock when what sounded like the faint drone of an engine pierced the hush. Not the cadence of a helicopter, but the steady hum of a fixed-wing headed somewhere through the night. Reflexively staring in the sound's direction, I strained to determine if it might be drawing closer. Nothing marred the cloudless sky except a low-riding moon and a multitude of stars that coalesced in one brilliant swath as the Milky Way. *Mmmm, Milky Way.* My stomach clutched at the thought. The hum faded away. I didn't belabor the possibilities.

I turned back toward the Big Dipper, my guidepost to guard against a haphazard choice at occasional two-track intersections. Pulling my flashlight from a cargo pocket, I rechecked my progress on the Crow Mountain map. I reserved the light for periodic inspections of the map—just in case I should need it later for unforeseen circumstances. Buoyed that barely three miles remained to my destination, I stuffed the map and flashlight back into my parka. Then I heard it again—the mounting drone of the engine.

As the resonance swelled, the sky burst into crimson light, and the trailing glow of a flare floated to earth. The craft audibly responded, though I saw only a blinking green light several miles to the east. My face flushed with emotion.

"They made it," I mouthed. Rawley and John were OK.

I felt momentarily conflicted. I was far closer to the Duncan place, but felt drawn backward toward the cabin. *No. Stay with the plan.* The airplane merely established our location. A helicopter or a search party on snowmobiles was needed to reach the remote cabin. I pushed ahead with renewed spring in my legs.

🦌 🦌 🦌 🦌 🦌

Well after dark, Rawley and John reached the cabin. John was all in. Wheezing, pale, and wracked by leg spasms, he tumbled onto the bare coil spring bed positioned along one wall of the 10 x 10 foot shack. There he sprawled, kneading his legs, while Rawley inspected the wood-burning stove. Finding no firewood inside, he went to the livestock corral and tore loose several of the thinnest fence rails. He broke them in pieces and started a fire. Now more than ever, he considered John a heart attack waiting to happen. If no rescuers arrived by daylight, he planned to stoke the fire and follow my tracks to the Duncan place.

From inside, Rawley heard the whisper of an engine. He bolted through the door. John joined him outside and produced the red flare gun from his jacket. Minutes after the night sky above the cabin exploded crimson, the Cessna buzzed the cabin and wagged its wings. John knew that within moments a helicopter from the home base in Greybull would be dispatched to rescue them.

An hour later, the "thwack, thwack, thwack" of changing rotor pitch echoed off Trail Ridge. A searchlight advanced, then panned about where I guessed the cowboy cabin rested. The light glided downward until extinguished by the low ridge I had just descended.

It was after eight o'clock when the cadence of rotor blades returned. A searchlight traced my footsteps. Shortly after, I climbed through a rear door of the Jet Ranger and was reunited with the others. As we lifted skyward, an amber yard light glimmered a mile west at the Duncan place. I had been hiking for nearly eight hours, and felt like it.

We were back at Thermopolis in 20 minutes, as if nothing out of the ordinary had happened. Nothing, that is, except a stern reproach from both pilot and copilot that we should have remained with the Hiller. The ELT's signal had activated. The Cessna had searched for us there first, then had spotted the flare at the cabin. While their efforts to find us were mighty welcome, none of us was in the mood for the ass-chewing we received.

"What were you thinking?" Dan Hawkins scolded. "You of all people should know to stay with the helicopter, John!"

Well, personally I was thinking warmth, food, water—maybe survival—all of which I was on my way to securing had a helicopter rescue not occurred that night. But no need to prolong the chitchat, I thought. All's well that ends well. Besides, I felt too exhausted to argue the point. *But next time, I will make sure there is survival gear on board an H&P helicopter. What about that, guys?*

6

The Way It Was

I do not want to settle down in the houses you would build for us. I love to roam over the wild prairie. There I am free and happy. When we sit down, we grow pale and die.

—Chief White Bear, Kiowa

Traveling the reservation, I found the Christian mission church remained the most prominent building in many small towns: St. Edward's, St. Michael's, St. Stephen's. Those stone and log chambers were where God-fearing men and women found the Great Spirit, rather than in the peaks and plains, rivers and woodlands, or in the eagle and bison. The incongruity was no less poignant than announced by a motel I once passed on Route 66 near San Bernardino, California. The motel consisted of 19 stucco tepees. The sign in front read, "Sleep In A Wigwam—Get More For Your Wampum."

A half mile beyond the gravestone of Sacagawea and another two miles southwest of Chief Washakie's resting place stand the log buildings of Father John Roberts' mission, the reservation's first, built in 1887. For 55 years in the adjacent Shoshone Episcopal Mission Boarding School, on lands consecrated by tribal rituals, Father Roberts taught Indian girls their Christian catechism. To some, these juxtapositions symbolize how missionaries brought the Word of Christ and a proper education to Native Americans. To others, it is a reminder of how waves of Spanish, French, and English intruders conquered, subjugated, and culturally cleansed hundreds of American nations.

These incursions began with the conquistador Hernán Cortés in the sixteenth century, followed by James Smith, the Pilgrims, and a flood of others a century later. Whether they were plundering and subjugating the Aztecs, enslaving Africans to furnish cheap labor for America's plantation economy, or defeating and displacing indigenous Americans from their homelands, European and Euro-American conquerors have often viewed aboriginal peoples and their cultures as unworthy of respect and have treated them as a convenient conduit to wealth or an obstacle to geographic expansion.

American Indians were expected to summarily assimilate and accept an alien culture and way of life. This callous imperative bore untold casualties along the way. Can anyone look upon the Cherokee Trail of Tears, Wounded Knee, or the Nez Perce's three-month retreat from the relentless pursuit of 2,000 cavalry and not reflexively sense their loss and feel a consequent pang of remorse? Today, news of a formerly "undiscovered" tribe in New Guinea or the Amazon still compels me to shout, "Leave them alone!" Who has a corner on the market of contentment and well-being?

The Nez Perce Chief Joseph encapsulated the matter this way, "The earth is the mother of all people, and all people should have equal rights upon it. You might as well expect the rivers to run backward as that any man who was born free should be contented when penned up and denied liberty to go where he pleases." Ironically, the U.S. Constitution embodies this same spirit.

A century later, I sometimes found the outcome heartbreaking. The Wind River's debilitating rates of diabetes and alcoholism and demoralizing unemployment overwhelmed me at times. While some impulsively attribute this to an indolent people, such stereotyping expresses continuing prejudice and naïveté. I saw it differently. Learning their history steeled my resolve to do all I could to restore part of the Shoshone and Arapaho peoples' heritage, their wildlife.

Not unlike white missionaries, miners, and fortune seekers—cultural refugees displaced by an ocean—the Shoshone and Arapaho nations were civilizations defined by their pasts. It struck me that I couldn't chart a future for WRIR's wildlife if I didn't understand its condition in former times. That is, I couldn't formulate a revival of past game herds without piecing together their histories. Past, present, and future were equally relevant ingredients of a conservation formula.

I assumed that the WRIR could support larger populations of big game. But was that true? A fatal mistake a scientist can make is to assume he knows the answer to a problem not yet investigated. If my goal was to advocate optimum populations of big game, whatever that optimum might be, I needed to plot a course for population change. As my former college professor Les Pengelly once noted at a science conference I attended, "You can't get to the North Pole if you don't know which way is north."

I discussed my ideas first with Dick, then with Wes Martel and Frank Enos, Shoshone tribal councilmen who always gave me good counsel. Wes's weathered face betrayed a man scarcely older than me. His craggy features and graveled voice masked a quiet intensity and spiritual union with the Wind River's wildlife, people, and land. Our conversations about the latest habitat statistics or wildlife trends served as cultural lessons to me. He spoke passionately about his kinship with his surroundings and of the role that all things both living and inanimate played in traditional Indian life. A perfect counterpart as my reservation mentor, Frank's approach was more practical. His veterinary science training matched my pragmatic approach to our common purpose. When I needed feedback from a trained ear or an ally to prime the tribal council, Frank was the sounding board and advocate I sought. These men's leadership and trust in me steered the wind to my back in times when the uncertain course of conservation's waters grew rough.

I added their suggestions to my list of tribal elders that Dick suggested I visit. New contacts in turn suggested others I should meet. Each proved another book in the library of knowledge chronicled as memories of their life experiences. That knowledge was not centrally organized and gathering dust on the shelves of some austere building. It was alive; dwelling in communities, homes, and tepees scattered across a million or more acres.

In 1979 and '80, I interviewed twenty-four longtime reservation residents ranging from 59 to 89 years of age: thirteen Shoshones, six Arapahos, and five non-Indians who lived on reservation inholdings. The latter group included longtime ranchers owning fee-patent lands. Important points and deductions I confirmed from two or more sources. From these absorbing conversations, compelling patterns emerged.

🦌 🦌 🦌 🦌 🦌

In the Owl Creek Mountains, 90 percent of all elk I observed from aircraft were restricted to the most rugged and forested habitat in the western third of the range. Habitat in the eastern Owl Creeks was certainly suitable for elk and other big game, but roads now crisscrossed the rolling topography, providing easy access to hunters much of the year. Sound or sight of any vehicle, including mine, sent gun-shy animals scattering for cover.

But had wildlife always been scarce there? If so, why? If not, when had numbers waned? I would soon find out.

The Dukarika or "Sheepeater" Indians were among the earliest inhabitants of what is now WRIR. They were kin of the Eastern Shoshone Tribe and dwelt in mountain retreats, including both ranges of WRIR. They hunted primarily bighorn sheep, but also deer, elk, and rabbits for food and hides.[1]

The Arapahos largely inhabited lands to the south and east of the present WRIR both in Wyoming and in adjacent states. By 1800, other more powerful Plains tribes had driven the Eastern Shoshones west of the Wind River Range.[2] Although hunting grounds to the west and south satisfied most of their needs, they continued to visit the "Warm Valley" (what Shoshone people called the Wind River Basin) on spring and fall buffalo hunts—more so after 1840 when buffalo became scarce west of the Wind River Range. To supplement dried buffalo meat when winter supplies ran low, Shoshones and Arapahos hunted deer, pronghorns, elk, and other game.[1] Like a continental tsunami in the late 1800s, Euro-Americans decimated these animals, the lifeblood of these and other Plains tribes.

During the twentieth century, game populations fluctuated with changes in land status, hunting pressure, and accessibility of habitats. In 1904, that portion of WRIR north of the Wind River, nearly 1.5 million acres, was ceded from the tribes and opened to homesteading.[3] Following that action by the U.S. government, hunting was regulated by the state of Wyoming with licenses required for Indians and non-Indians alike. During 21 of the 35 years that the northern lands were ceded, the State closed the hunting season on the depleted pronghorn population, as it did throughout most of Wyoming. By then, a continental population estimated at 40 million had dwindled to 13,000. Unrestricted tribal hunting of pronghorns and other wildlife continued on the diminished reservation, a prerogative under tribal treaty rights.[4]

Pius Moss was an Arapaho elder. At 66, he told me he had spent every day of his life on the reservation. His furrowed brown face—resembling the dendritic pattern of the Wind River badlands—suggested many were spent in the sun and the wind. Seated at his kitchen table, I asked him with what parts of the reservation he was most familiar. He replied without sarcasm, "Those parts from east to west."

According to Pius, deer all but vanished from the diminished reservation but became plentiful on the ceded area. "It was easy to harvest a deer almost anywhere in the Owl Creeks."

"Why," I asked, "the difference between the two areas?"

"The deer increased on the ceded lands after the State controlled the hunting."[5]

I found this statement astounding. Here the U.S. government had seized over half of tribal lands, and this Arapaho gentleman graciously acknowledged a service rendered by another governmental authority. But he wasn't alone.

Pearly "Junior" Brooks and Shoshones Bill Bradford, Sr., and Jim Barquin, all longtime cattle ranchers in the Owl Creek Mountains, attested the abundance of pronghorns, deer, and elk there during the 1930s and early '40s. Simply put, Arapaho Jess Miller said pronghorns "numbered in the hundreds and hundreds in the Wind River Basin."[4]

"They were thick from Fort Washakie to the Owl Creek Mountains," is the way that Herman Lajeunesse described pronghorn numbers 40 years prior to our meeting in 1980. In my most memorable interview, Herman provided me a wealth of information on reservation wildlife. As his wife, Rachael, served us coffee and homemade desserts, the three of us talked and laughed for two hours. Soon I forgot

Herman and Rachael Lajeunesse in 1965

I was in a stranger's house; rather it was a home full of life and fond memories. Taxidermy trophies of deer, elk, pronghorns, and bobcat roamed the walls. A black bear hide, complete with a gaping-mouthed head, sprawled across the couch next to the chair where I sat. Outnumbering the animals that peered through glass eyes were high school graduation pictures of Herman and Rachael's children. Thirteen of fourteen had received high school diplomas—an amazing accomplishment for any parents. Their cap and gown portraits and basketball trophies crowded into a glass-fronted hutch and showed the pride of their parents.

Herman was 80 at the time. With quick eyes and a quicker mind, he told me how he had guided the U.S. Geological Survey party that mapped the Wind River Range during the 1930s. "I packed in their food and equipment and showed those boys how to get around."

"The entire range?" I marveled.

"From Dinwoody to Little Wind, and every pond and knob between," he chuckled.

During the 1940s, Herman became only the second game warden to work on the WRIR. His experience afield, as a trapper, warden, hunting guide, and ardent hunter himself, surpassed that of anyone else with whom I spoke. His great-nephew, Gary Lajeunesse, likewise became a tribal game warden, and a friend of mine during my reservation years.

Regarding elk, numbers were critically low throughout Wyoming in the early twentieth century. Northwest of WRIR in Yellowstone National Park and Jackson Hole, most of the continent's remaining elk survived, reduced from millions to 50,000. Remnants likely persisted in the reservation's western mountains on the outskirts of the Greater Yellowstone Ecosystem, an 18-million-acre tapestry of rugged ranges. Buried beneath snow for half the year, these wildlands were sufficiently remote for elk and grizzly bears to make a last stand.

Herman claimed that the only resident elk herd in the reservation's Wind River Range during the 1920s and '30s was a group of about 25 in Bull Lake Canyon—perhaps the most rugged terrain on WRIR. An Arapaho story provides testament to the mystical significance of that place. During the nineteenth and twentieth centuries, the Arapahos believed that the buffalo were chased back into a hole in the ground by the white man. The Northern Arapahos say that this occurred somewhere in Wyoming, with one account naming the area as Bull Lake. We now know better—or do we? On my first hike into that canyon I came upon two weathered bison skulls. Both rested not far from a dark hole in an outcrop.

Elk were also scarce in the Owl Creeks, so scarce that Wyoming supplemented the remnants with 11 elk shipped from Jackson Hole. They were transplanted west of Wind River Canyon in 1918. Buoyed by State-imposed hunting regulations, numbers increased. By 1930, State Game Warden Cal King wrote that groups numbering as many as 75 could be seen in the eastern Owl Creeks.[6] Reservation residents Pius Moss and Jim Hill (in whose cabin John and Rawley took refuge on our ill-fated flight a year earlier) told me, "It was common to see groups of 100 or more elk throughout the Owl Creeks during the 1930s, because there was regulated hunting."[7]

Both afoot and from the air, the Owl Creek Mountains impressed me as primo elk habitat. I understood why the eastern Owl Creeks might support few elk in summer and fall, given the lure of extensive high wilderness to the reservation's northwest. Most Rocky Mountain elk herds are migratory, spending July through September at high elevations grazing the lushest vegetation. They convert grass and herbs to milk for their young and to muscle and fat in preparation for the rut and for winter—times of high energy demand.

From elk wearing neckbands attached by State biologists west of the reservation, I found that Owl Creek elk migrated 10 to 20 miles from the DuNoir and Washakie wildernesses—fabled elk summer haunts—to the western Owl

Bighorn sheep in the Wind River Mountains, December 1978

Creeks in late fall. Elk could have migrated another 20 to 40 miles farther east, but they didn't, at least not now. This would have opened up 150 square miles of additional winter range; bountiful grasslands their precursors grazed.

I assumed that bighorns once managed that same migration. At the eastern terminus of the Owl Creek Mountains, the safe haven of Wind River Canyon featured the kind of ledgework and breaks that are magnets to sheep. Prime territory also for the student of geology, the canyon's walls dive 2,000 feet to the grinding river below. As the Wind River slices through the earth's outer crust, it peels away eons like layers from an onion. The oldest Precambrian rocks date to 2.9 billion years before the present; before the dawning of slime molds and segmented worms. Driving State Highway 20 north through the fourteen-mile-long canyon, time flashes forward to the onion's outer strata, dating a youthful 208 to 245 million years ago. These red Chugwater shales of the Triassic Period were trod by the earliest sauropod and theropod dinosaurs.

Wild sheep readily graze ridges and plateaus so long as the safety of steep terrain is handy. Although the eastern Owl Creeks largely lack such security, sporadic outcrops and south-facing escarpments seemed suitable for hopscotching west to east. Early explorers' reports of "plentiful mountain sheep" in the region substantiated my academic view, as did the recollections of those I interviewed.

Across a map spread on his table, Art Nipper, Jr., a Shoshone, traced how bighorns still traversed the Owl Creek chain in the early 1900s. I met Art at his house on lower Crow Creek where Black Mountain loomed provocatively to the north. Linking the map to the northern horizon, he indexed the places where bighorns made their winter homes, including Black Mountain's craggy slopes and Crow Mountain, its little twin to the west.

"Yes, there and the red sandstone badlands to the south are where Jim Fike and Burke Johnson saw bighorns just ten years ago," I related.

"Maybe there are still some left." His eyes beamed widely.

"I don't know," I replied softly. "I've hiked and flown over there, but haven't seen any bighorns. Only weathered skulls and horns."

His expression faded to disappointment. But as his eyes and a crooked finger found the map again, his thoughts returned to lost migrations of the 1920s. "Phlox Mountain and Sheep Creek," he pointed. This and other rough country linked Washakie Wilderness and Wind River Canyon. "Then across Mexican Pass to the [Wind River] Canyon where I would see some from December to April." He watched the sheep return west in May and June, bearing bouncy lambs on their way to alpine summer ranges.[8]

As often happened, Art's recollections were bolstered by others, extending the timeline of the bighorns' waning migrations at least a decade or two. Ernest SunRhodes, an Arapaho, also recalled bighorns in the gentle country near Mexican Pass, just 10 miles west of Wind River Canyon.

"Bighorns lived on both sides of the canyon. Sometimes they crossed the river and highway," stated Landis Webber, who saw a Trailways bus plow into one on Highway 20 in 1949. Martha Stagner recalled that as recently as the summer of 1964, a friend had observed a solitary bighorn scrambling the canyon's cliffs.[8]

Since 1964, only Webber, who ranched in the Owl Creeks, recalled sheep well east of their present distribution. "Over three weeks in winter 1979, I watched a band of twenty-five on Phlox Mountain."

There the trail goes cold. Those last pioneering sheep apparently died out, and with them the herd memory of Phlox Mountain, Sheep Creek, Mexican Pass, and Wind River Canyon. But more than the Owl Creeks, the Wind River Range had once been the stronghold of the most abundant big game animal on the reservation.

"There were bighorn all over the face of the Wind Rivers in the 1870s," Charlie Appleby of Lander had told his young friend Carl Shatto, who in turn relayed it to me.

Burke Johnson confirmed their abundance following WWI this way: "On horseback, my dad had no problem shooting a bighorn in one day whenever he wanted."

The demise of bighorns was even more dramatic in the Wind River Range than in the Owl Creeks. Burke Johnson and Art Nipper told me that bighorns crashed at the height of domestic sheep grazing from 1927 to 1938. "In the 1920s, there were thirty-five sheep operators in Lander alone herding about 325,000 sheep across the Wind Rivers' slopes," Carl Shatto recounted. Another 100,000 were grazed in

Bighorn winter range in the Wind River Mountains, December 1979

the Owl Creeks, according to Shoshone Tribal Chairman Bob Harris, Sr.[8]

This portrait corresponds with precipitous declines of bighorns throughout Wyoming and the West beginning in the late 1800s. "Too much" was the reason—too much hunting and livestock. Market hunting by Euro-Americans, uncontrolled sport hunting, and competing livestock initially extirpated sheep from the more accessible portions of their range along river breaks, foothills, and isolated buttes and mountains. Bighorns remained abundant on more remote ranges until domestic sheep overran even those.[9] Some of these "prairie lice," reviled by many cattlemen, were trailed from as far away as Oregon to graze the lush pastures of Wyoming's Wind River Range.

Like a plague, domestic sheep walloped bighorn ranges. They competed with wild sheep for food and space. What is more, they transmitted scab mites, bacterial and viral pneumonia, and other diseases to which bighorns were highly susceptible. Like American Indians lacking resistance to smallpox and other Anglo diseases, bighorn numbers had collapsed. In this respect, disease-prone bighorns suffered more than other big game at the hands of the white man and the blight of his livestock.[9, 10] Our inventories of big game winter habitats from 1979 to 1981 found bitterbrush plants—two-foot-tall shrubs relished by wild and domestic herbivores—still misshapen like pampered bonsai. I have seen such damage to bitterbrush only where domestic sheep have grazed game ranges by the thousands. Despite removal of all nontribal livestock in 1940, bighorn numbers failed to rebound.

༄ ༄ ༄ ༄ ༄

Elk, deer, pronghorns, and bighorn sheep were each distinctively valued by the tribal members I interviewed. Historically, all five species (including both whitetails and mule deer) provided food, clothing, and tools and were admired for their grace and survival adaptations. Bison had long been a favored food of the Arapaho and Shoshone tribes, and bison hides draped on lodgepole frames sheltered the people and provided bedding at night. However, the supple yet durable leather from elk and deer hides clothed these peoples as well as fed them for generations. When the buffalo migrated beyond a clan's hunting grounds, the more sedentary deer remained to hunt in the basins and foothills. In Arapaho mythology, deer (*bíh'ih* in Arapaho) were associated with women and romantic desire. Following bison, elk provided the next largest protein source. The cultural importance of elk dates as far back as the Anasazi, who carved elk petroglyphs into the Southwest's sandstones centuries ago. As when hunting deer, hunters used whistles carved from bone or wood to imitate the call of elk during the stalk.

Moose were a recent arrival to WRIR, and to Wyoming in general. The Indians' ancestors apparently found the species a most curious conversation piece. I recall Herman Lajeunesse's response to my inquiry about historic moose numbers on the reservation.

"I was just a kid then," he noted, reminding me he was conceived in the previous century. "The first one that showed up on the reservation, why, nobody knew what it was!"

Yes, I could imagine the incredulous reactions. Some bizarre horse? A four-legged Bigfoot? Or—that's a lot of meat!

Every canyon in the Wind Rivers offered suitable habitat for moose, yet my flights produced pitifully few. Cottonwood and willow stringers dissecting Wind River Basin were also quality range, yet were missing moose. My interviews garnered scattered accounts of singles and pairs along those riparian arteries, but most ended with the moose made into meat. Moose were just too obvious and hopelessly vulnerable. Given that they're also quite tasty, I marveled that one occasionally wandered as far east as the towns of Arapahoe and Riverton before receiving its *coup de grâce*.

I observed only one moose per three hours of flight time over the Owl Creek Mountains. Although inferior habitat to the Wind Rivers, this range offered moose sizable patches of aspen, firs, and willows throughout its length. The only moose seen in recent years at the eastern margin of the Owl Creek Mountains was recorded by Frank Baggs of the Wyoming Game and Fish Department. In May 1975, he spotted a female in Wind River Canyon, surely a traffic-halting sight. It wore a numbered metal tag in one ear. This piqued Frank's interest, given that no moose was ever captured and tagged in the area.[11] His check of Department records showed she had been tagged near the Gros Ventre River, some 125 air miles west! Through the convolutions of canyons and conifer forest, she had seen more country than most moose ever do.

🦌 🦌 🦌 🦌 🦌

In 1939, the ceded lands were restored to WRIR. Restored, that is, minus a wedge of 400 square miles of potentially irrigable land, dubbed the Riverton Reclamation Project. As with past land withdrawals, the government claimed lands valued by non-Indians—in this case for farming and town sites. With the return of these lands to the tribes came the revival of unregulated Indian hunting. Deer, elk, and pronghorn numbers declined once again.[4, 5, 7]

The elk population in the Wind River Range increased markedly south of the reservation line during the 1940s as a result of stricter state hunting regulations. But Herman Lajeunesse, who was a tribal game warden at the time, said, "Elk numbers remained low on the reservation side until a tribal game code was passed in 1948. After that, elk moved onto the reservation from the south and north." They established a sizeable elk herd in the roadless area, then spread throughout the entire range during the 1950s.[7]

The two tribes adopted the 1948 game code to stem overexploitation of wildlife. The code was strict. It provided for a September through December hunting season for only one bull elk and one buck deer per hunter. It prohibited all hunting of pronghorns, bighorns, moose, and upland game birds. The curbs on hunting, along with the presence of fewer hunters during WWII and the Korean War, fostered a resurgence in all game species.

By the mid-1950s, Jim Barquin recalled seeing "400–500 elk near Spring Mountain" in the Owl Creeks, and Pearly Brooks recalled "100–150 deer every evening in my hayfields" north of the Owl Creeks in the 1960s. Few remained 10 to 20 years later.[5, 7]

As we sipped coffee at his home north of Arapahoe, Jess Miller reminisced how he regularly saw deer among his pastured cows along the Wind River during the 1950s, a rarity anymore. As his fingers combed through his shock of silver hair, he gazed wistfully out the kitchen window. "I saw a group of whitetails out there last spring. I planned not to tell anyone. But a neighbor found out and shot at least one." He sighed and added, "I haven't seen any since."

With searing conviction, this past tribal chairman counseled me on the need for a game code. He feared that reservation game now lived on borrowed time. "Our young ones don't have the respect for wildlife," he lamented. "They just don't know the old ways."

The 1948 game code was short-lived. The tribes rescinded it in 1953, primarily because it was unnecessarily restrictive and lacked flexibility to make annual adjustments. Only three regulations were retained: prohibitions on selling or trading game meat, hunting with artificial lights at night, and wanton waste of game animals. Even these were spottily enforced.

The most recent decline of big game began in the 1960s, but largely occurred in the 10 years before my arrival. This was a unanimous opinion among those I interviewed. Shoshones Fred Harris and Bud LeClair, Sr., believed that elk had dwindled since the 1950s even in the rugged recesses of the Wind River Range.[7] They were vulnerable because their winter ranges shadowed the mountains' more accessible eastern flank.

Arapahos and Shoshones agreed: a fatal brew of factors precipitated wildlife's retreat. Increasing numbers of roads and 4 x 4 vehicles, like the chicken and egg, made access easy where once only horses were ridden. High-powered rifles replaced traditional weapons. Coyote predation, poaching by nonenrolled persons, and snowmobile harassment were cited as secondary reasons by some. With the development of oil and natural gas resources on tribal lands in the 1940s, enrolled members shared profits in the form of monthly per capita payments.[2] That reliable income provided members more leisure time, meaning more time to hunt than in the past. With 12-month seasons and bag limits constrained only by the size of one's freezer, the future for wildlife was bleak indeed.

The roller-coaster ride of reservation game's well-being, described by those who experienced that ebb and flow, clearly illustrates how contemporary technology and incompatible behavior begot wildlife's ruin. Of course, the worsening situation was why the tribes had asked for help. That was why I was there. And soon a reckoning would arrive. Tribal members would have to choose—continue unfettered hunting, or restrict their sovereign treaty rights.

🦌 🦌 🦌 🦌 🦌

Now that three decades have passed, so have Herman, Jess, and nearly all the rest. I consider those interviews—the historical record they produced and the lilt of the conversations—among my most memorable experiences of that time. As I listened to these descendants of the buffalo economy, their stories sometimes strayed unpredictably. Old men's eyes occasionally looked beyond me as they spoke of loved ones and the land, and of things I would not have thought to ask, and some I could never understand. I watched their furrowed, copper-skinned faces and tried to comprehend their connection to the animals of which they spoke. Perhaps some found solace still living in this ancestral place; a biological, geographic, and spiritual continuum trailed a hundred centuries into the past. The sun still rose above the eastern plains where brother buffalo had bounteously roamed. The stars and moonlight bathed the continental spine where glacial flows marked their people's dawn in time.

Sharing their coffee and stories, some wore braided hair. Some wore hats with John Deere or Redman Tobacco logos. All were grandchildren of a proud and fading culture. With their consent, a tape recorder whirred. It freed me from the preoccupation of scribbling salient details on notepads. As the recorder chronicled unwritten history, I too succumbed to moments of mind-drift. Visions materialized of what I pictured once was.

I saw elk and deer dotting swells of sagebrush and colorfully banded badlands. Herds meandered the windswept contours of the Owl Creeks, which also teemed with buffalo, grizzly, and grouse. Bighorns roamed every canyon and bench of the Winds. Their dusky brown bodies, bookended by suntan flared horns and ashen rumps, bounded wherever gravity allowed. In the basin of the Big Wind, hosts of 20 and more tan and white prairie ghosts drifted across stippled sagelands. And there beside Crow Creek and the Wind and Popo Agie rivers, dozens of buffalo-wrapped pyramids graced the meadows. Gray columns curled from the apex of lodges and cook fires as the people fleshed hides, dried meat, knapped chert, and flourished beneath sun and stars.

These panoramas predated the experiences of those I interviewed. They were products of grand orations and daily instruction passed to successive generations as surely as bound volumes of Shakespeare and Plato. The memories I tapped encapsulated the transition between a lifestyle lost and a world thrust upon the Plains Indians by Euro-Americans. It was a time of turmoil for populations—both native animals and native peoples. Fifty years, spanning the late 1800s and early 1900s, altered the western frontier forever and nearly ended the living record of the large mammal populations I now sought to reconstruct. As common victims of western settlement, remnant populations of buffalo, elk, and Indians were treated similarly. Parks and refuges were designated to salvage the animals, reservations to sequester Indians.

For nearly two million years, the Pleistocene glaciations advanced, receded, and swept again across the Northern Hemisphere, wreaking havoc on flora and fauna. During a relative blink in time, the Euro-American advance on the American West was selectively repeating the same. This time it progressed at the hands of a conqueror come to watch. Over 13,000 years earlier, Clovis culture—the mother ancestry of American Indians—had also transformed the landscape forever. Clovis hunters served as principal agents (arguably partnered with widespread habitat change) in the elimination of Pleistocene megafauna—mammoths, mastodons, short-faced bears, and saber-toothed cats. Elk and other extant species, which radiated and filled vacant niches, became the Holocene beneficiaries. These novel newcomers, unlike their vanquished forerunners, had coevolved with protohuman hunters in Eurasia. They had adapted over millennia to the upright huntsman's prowess. Then a new wave of New World humans with another mentality and lethal technology nearly swept them all away.

In the view of aboriginal people, everything is connected. All life interacts on a horizontal plane. People's thoughts, feelings, actions, indeed their very existence, is interwoven with Earth's other creatures and the land that sustains them. With no hierarchy of importance elevating one animal above others, including the human animal, no place exists for feelings of arrogance or subservience. Animals possess unique traits that are admired and even worshipped by people who learn not *about* animals, but from them.

This respect toward the rest of Nature is common among indigenous peoples on every continent. As an example, I recently traveled to Africa where a native guide in Botswana, clad in khakis and using the English name Dan, held fast to his traditional belief system. On our first morning's game drive, we passed a foraging troop of chacma baboons. "Those are my cousins," he matter of factly declared.

I and the others in the Land Rover chuckled at the joke. The next time he remarked likewise about these communal primates, I listened to the tone of his voice and knew better. He was sincere.

Because of the Euro-American worldview—ordered vertically with our dominion unquestioned at the top—it seldom occurs to us that we might learn from other cultures. This also helps explain why the Western concept of ownership of land and what lives there is such a foreign concept to aboriginal people. The land has boundaries to those people no more than it does to an eagle or an elk. Obviously, much has changed. Wherever indigenous people have felt the hand of European oppression and influence, their traditional views have become variously corrupted. That makes their cultural heritage no less relevant to their lives and no less worthy of our respect, in my view.

I'm not suggesting that aboriginal Americans lived in idyllic harmony among mammal, bird, and fish without impact. That is simply not the history of humankind's global imprint, as Jared Diamond details in *Collapse*.[12] However, with primitive technologies, a hunter-gatherer way of life, lower population numbers, and cultures grounded in Nature's mysticism, early Americans and Holocene wildlife coexisted and prospered for 100 centuries. Then everything changed. Confined to postage-stamp homelands and armed with contemporary technology, Indians were subsidiary participants in the plight of wildlife as Manifest Destiny forever altered the American landscape. Now, with an enlightened knowledge of our hegemonic ways, conservation calls us all to the cause of restoration.

Armed with hours of these people's testimonials, I reconstructed a century of history that preceded my arrival. To share with all Shoshones and Arapahos, I transcribed the saga of the WRIR's wildlife in their peoples' own words. I hoped it would serve as an important cog in a wheel to recovery. If change came, it would rise from the raised hands of those living on the WRIR. Still, I wanted one testament more to strengthen the case.

In 1980, the Tribal Fish and Game Committee and I designed a questionnaire on wildlife matters. The Joint Business Council printed the survey on its letterhead and mailed it to 3,818 enrolled members of the Shoshone and Arapaho tribes, all those 14 years of age and older.

One question asked respondents to rate the importance of each big game species on WRIR. Another asked which they preferred to hunt. Another listed resident and migratory bird species and asked which of these they hunted.

The most general question asked whether big game populations on WRIR were smaller, the same, or larger than in the 1950s. Of the 618 Shoshones and 583 Arapahos who responded, 85 percent of Shoshones and 72 percent of Arapahos said big game numbers were smaller than in the 1950s. I was bowled over! This heralded great hope that both tribes' members were ready for change. As rock-solid evidence of the importance of wildlife, 79 percent of Shoshones and 89 percent of Arapahos said they hunted big game. Even more surprising, Indian women said they hunted big game in significant numbers—67 percent of Shoshones, 83 percent of Arapahos.[13] I took this endorsement as support for restoration of wildlife. But did they connect the dots as I did, with the path to sustainable wildlife and hunting requiring self-imposed curbs on the latter?

In a larger context, the questionnaire results were even more striking. A 2001 survey of the American public showed that only 6 percent of us hunted *at all*—a steadily declining trend over recent years.[14] In Wyoming, a wildlife-rich and rural state where I would expect higher participation, just 17 percent of those 16 years of age and older hunted, according to a 2002 survey by the National Shooting Sports Foundation.[15]

I knew that this survey of the Wind River tribes was no Gallup poll product reeking with statistical levels of probability. I didn't verify whether the attitudes of nonrespondents mimicked those of the people who returned the survey. I couldn't estimate the margin of error associated with the results. But all surveys are fraught with biases, even those conducted by highly paid pollsters. Nonetheless, at face value, the responses we received showed exceptional interest, if not genuine participation, in hunting. Even if I assumed that *everyone* who didn't respond—61percent of Shoshones and 74 percent of Arapahos—didn't hunt, that would still leave a hunting participation rate of 22 percent. Among all 50 states, that would trail only Montana at 24 percent.

Surveys of the American public at large show that those living in rural areas hunt more than do urbanites. The entire WRIR is rural. Hunting is a lifestyle of its people tied to place, just as attending the theater and strolling Central Park are lifestyle pursuits of New Yorkers. Add in the traditional importance of wildlife to the tribes, and their freedom to hunt whenever they chose, and the survey's results were compelling. Now to put these revelations to work.

7

Younger Kids

You can teach a student a lesson for a day; but if you can teach him to learn by creating curiosity, he will continue the learning process as long as he lives.

—Clay P. Bedford

I was making progress divining the status of reservation wildlife. My interviews of elders painted a portrait of the past. Comparing past with present would show how things had changed, but would still not reveal what could be. "Could be" concerned potential—future populations that available habitat might sustain. Perhaps they were numbers never experienced by living Arapahos and Shoshones. I would trust two lines of evidence to make these predictions: comparisons with wildlife densities adjacent to the reservation, and carrying capacity estimates based on our winter range forage inventories.

My biggest challenge at WRIR would be persuading tribal members that change was needed—change in hunting behavior, foremost. How would sacrificing some of their hunting freedom benefit Shoshones and Arapahos? What would be the consequence of continuing the status quo?

That was the case I needed to make. A tall order it seemed, indeed.

But why, I stewed, should Arapahos and Shoshones believe or trust me? Especially with something as entwined with their lives and culture as wildlife? A sordid history of breached treaties and broken promises by other government officials was unforgotten. Maybe unforgiven.

I knew I had no ulterior motives. I wanted nothing from these people. Still, I needed to earn their trust and present no reason to fear betrayal. Most of us, regardless of our ethnicity, require as much. These internal conversations plagued my thinking. They also drove me to find ways to bridge the cultural divide.

The clock was already ticking. Our Technical Assistance station's supervisors in Billings, Montana, assured me and project leader Dick Baldes that I would be given sufficient time to complete a comprehensive wildlife management plan for WRIR. That was my paramount task. Then they added the caveat. All my work, including the plan, should be finished in about four years. After that, I would be moved elsewhere.

🦌 🦌 🦌 🦌 🦌

I found that sincerity, persuasion, and education were tools as essential to my work as binoculars and helicopters. My informal conversations with Frank Enos, Jess Miller, and other councilmen and elders obliquely implied that the reservation's young people might not be instilled with a conservation ethic—at least not one compatible with restricting tribal members' hunting. Such values and moral standards in most societies are passed from parents or grandparents to children. Likewise in Arapaho and Shoshone cultures, familial education had traditionally passed acquired knowledge about the natural world to each new generation. So what had gone wrong?

I didn't have time to do a Ph.D. dissertation on the subject. Our office had begun discussing game code provisions with the Tribal Fish and Game Committee during my first months at WRIR. A reasonable game code based on sound science was one piece of the conservation puzzle. Informing and persuading the public that hunting restrictions were in their best interest was an equally important piece. A reservation-wide message connecting wildlife values, sustainability, habitat, and hunting was needed. If the message was redundant—information most tribal members already knew—it seemed at worst a harmless gamble. My ideas included a mailing to all reservation households, a newspaper column in the *Wind River Journal* (the reservation newspaper), even radio PSAs about our station's work. But I soon learned that in-school environmental education was an overarching need.

I wanted to reach as many people as possible, but I also had to be realistic about what I could take on. Marketing the message could become a time-consuming digression from my resource work.

"Don't spread yourself too thin," Dick cautioned me. "Writing a reservation management plan is going to be a big job. Don't get yourself sidetracked along the way."

Good counsel, I knew. There were just two other wildlife technical assistance biologists in our eight-state region of the U.S. Fish and Wildlife Service. Both did wildlife surveys and provided management advice to tribal governments upon request. But there were no integrated programs spanning comprehensive inventory, planning, management—and soon—environmental information and education. Ours at WRIR stood alone. Ours was becoming an ambitious program, to say the least.

I believed a successful management plan, and particularly tribal hunting regulations, would require broad-based buy-in. Otherwise, the plan might gather dust, and the voting population would reject a new game code. I pressed my supervisor, "If tribal members fail to see the necessity, there's no way they will limit their own freedom to hunt." Dick relented. "If it doesn't interfere with your primary work, then I guess it's OK. I'll support it."

I felt relieved as I walked back to my office to outline this new endeavor. *This could be the difference between success and failure for wildlife recovery.* Then it hit me. My 50-hour workweeks had grown yet again. Among a mix of outreach efforts, I soon learned that in-school environmental education would prove a worthy but frustratingly ambitious endeavor.

🦌 🦌 🦌 🦌 🦌

In his book *The Creation*, eminent Harvard scientist E. O. Wilson provides advice on how to raise a naturalist.[1] He emphasizes that becoming a naturalist is not like studying algebra or learning a foreign language. "It would be a mistake to introduce a child to Nature by a walk through a park or arboretum, with labels naming the species of trees and shrubs." Think of the child as a "hunter-gatherer" or "savage," he writes, savoring the virtue of the words. "He needs to thrill to the excitement of personal discovery, to mess around a lot and learn as much as possible on his own."

This admonition struck me as spot-on when I recall my own formative years. I wasn't pushed by others. Rather, my parents indulged my wild curiosity. Like others who grew up during the 1950s and '60s in rural America where lakes, marshes, fields, and forests were close at hand, I had the opportunity to explore Nature at my own pace. Mucking around in marshes, traipsing through woodlands, and assembling a collection of reptiles, amphibians, and assorted creepy crawlies opened Nature's doors to me. These early experiences sparked my interest in and pursuit of biology. But many if not most children today typically grow up much differently, watching television and movies, playing video games and sports, or hanging out at the mall. Electronic and structured activities have supplanted independent pursuits out of doors. While kids may own pets, wildlife and Nature are often distant and largely unknown to them. Living in this comparatively sterile and domesticated world, urban and suburban children are not just unacquainted with or apathetic to wild animals and Nature—they may fear them.

The growing disconnect between America's youngsters and the outdoors is well documented. Periodicals and books, like Richard Louv's *Last Child in the Woods: Saving Our Children from Nature-Deficit Disorder*, detail how kids now spend far more time indoors than did previous generations.[2] The allure of video entertainment, the decline of unstructured play, and increased legal liability are all to blame. The nation's growing culture of apprehension, as well as parents' fear of threats to unsupervised children, including "stranger danger" and Nature itself, have made the outdoors a stranger to our kids. Twenty-first-century Americans talk much about the environment but increasingly navigate Web sites, not rivers. As a consequence, conservationists predict waning citizen participation and advocacy for stewardship of wildlife habitat and conservation. The tribal elders I talked to (who served as spiritual leaders and grandparents but sometimes also as *de facto* parents of reservation children) were witnessing this when I arrived.

So why should people care about wildlife if they don't value it or understand its importance in a biosphere on which we ourselves depend? I think Suquamish Chief Seattle laconically captured the answer: "We did not weave the web of life; we are merely a strand in it. Whatever we do to the web, we do to ourselves." In modern cultures, this seems an increasingly unpracticed, foreign concept, an intangible ideal lost amidst societal trappings and economic expediency.

Aside from the conservation implications, a sedentary life indoors is by no means safe. Too much inactivity can lead to childhood depression, obesity, and attention deficit disorder. Youth experts report that kids need more unstructured play, which builds self-confidence and coping skills. It is up to parents to encourage outdoor time for kids, and also to be models. "You can't be outside with your kids if your life is too busy with other things. Kids watch us like hawks. We can't tell them to stop watching TV and talking on cell phones if that's what we're doing all day," warns Megan Ault, a Montana journalist and mother of three.[3]

In my estimation, reservation kids were headed down that road when I arrived. This should not seem surprising, despite their traditional ties to the land as a hunter-gatherer culture. A century of forced assimilation into Euro-American society stripped Arapahos, Shoshones, and hundreds of other Indian nations of a way of life that demanded knowledge of and engagement with Nature. Their lives instilled understanding and security in the outdoor environment that provided their needs and determined their survival. No Safeway or Walgreens lay just down the road. Once Indians were reservation bound in a wildlife-impoverished landscape and forced to depend on the U.S. government's terms of subsistence, the old ways began to fade. The generation of children I encountered was nearly as cola- and sitcom-addicted as their peers in Anytown, USA.

🦌 🦌 🦌 🦌 🦌

Our office provided nature films to teachers who requested them. But when I inquired, I learned that structured curricula in environmental science were lacking in reservation schools. I felt this was a missed opportunity—no, a necessity for fostering a conservation ethic among reservation youth. If kids weren't gaining an appreciation for the natural world from their families, then schools should be filling the learning void. As Senegalese ecologist Baba Dioum wrote, "We will conserve only what we love; we will love only what we understand; and we will understand only what we have been taught."

So I telephoned teachers, high school through elementary, at the four school systems on the reservation. I found teachers collectively interested in having me speak to their classes about my work on the reservation. Some hedged, however, when I asked if the *students* would be as eager to learn.

I began with a science class at Wyoming Indian High School in Ethete. They would be equipped to understand conservation concepts. They were also old enough—at least 14—to vote on General Council matters. The 20-plus students were polite as I described my work and how it could provide a more secure future for reservation wildlife. But many only fleetingly looked at me. None responded when I rhetorically asked, "Would you like more big game on the reservation to see and to hunt?" Indifferent, rather than polite, better describes their demeanor.

I moved on from the ensuing pregnant pause by concluding that the reservation's habitats could support larger populations. I added that each of them would have a say in that. After the teacher thanked me for coming, I asked if anyone had questions about what I had said. I looked out over 20-plus blank stares. Several gathered books together. Others fidgeted in their seats. Two boys in the back exchanged some words, then stared coldly at me.

"Thank you very much for listening," I said. "You live in a very special place with wonderful wildlife."

Driving back to Lander I thought, *that went over like a screen door in a submarine. Maybe education's not my thing.*

My next talk was scheduled a week later at St. Stephens Indian High School. Same subject, same engaging talk (or so I thought)—similar response to my first effort. *What was*

I doing wrong? I didn't consider myself a boring speaker. I tried to make it personal. *This should be an interesting topic to these teenagers,* I reasoned. I had expected a couple of questions, or some pointed comment, even if disagreeable—some rejoinder to begin a dialogue. Anything!

I had spoken earlier to the high school science class in Lander, a school where kids who were white like me largely composed the student body. There were questions and a lively discussion. The difference, of course, was that I was not *one of them* at reservation schools. They were shy, wary, likely suspicious of me, I rationalized. They had years of conditioning and reinforcement. I was reminded of the two Indian snipers I served with in Vietnam. We were brothers in arms at war, yet even under those binding circumstances we still struggled to cross the cultural divide.

I took the advice of a teacher friend: "Talk to younger kids."

My next two talks were to middle school classes. Ma-a-a-jor breakthrough. They were reserved, but I did get two or three timid questions.

As I went from the oldest to the youngest kids, the veil lifted. It was all about age and how rigid, or not, their thinking was; how inhibited, or not, their dispositions were. I spoke about my work on the reservation, explained more generally the aims of conservation and managing wildlife, and gave talks about endangered and threatened species—why some animals become scarce and how people could prevent this or remedy their plight.

As I progressed to the youngest students, I finally maxed out (or rather bottomed out) at the Arapahoe Elementary School's first grade class—my most memorable presentation. I told the teacher that I had never spoken to students that young, and I didn't have any children of my own for practice. So I asked for help with choosing a fitting subject.

We brainstormed on the telephone for a few minutes, finally arriving at the subject "Animals of the Forests and Mountains." I would show slides of local animals, talk about where they lived, what they ate, and some interesting facts about each. I wrestled some with the vocabulary I would use. Outside of my recent talks at schools, all my speaking experience had been to adult audiences—graduate school instruction, public meetings, scientific conferences. No first graders. *Let's see, these are six-year-olds. OK, I'll avoid all science jargon and clarify whenever their faces tell me that they haven't a clue what I've just said.*

I arrived at the classroom and found a slide screen erected in front of the chalkboard. At midroom, a Kodak projector was parked on an isolated table. Two dozen wooden desks tightly hugged the back wall. When the first grade teacher saw me pondering the gaping floor in between, she reassured me. "This is such a special occasion for the students that I invited both first grade classes—and the kindergartners," she added tentatively.

"Oh, OK."

"To make room, the children will all sit on the floor," she explained.

I surveyed the space, now envisioning dozens of miniature faces trained on me. *Kindergartners. Five-year-olds. What words do they know? Thank goodness for the slides!*

The kids filed in. First came the cutest bunch of six-year-olds I had ever seen, with glossy black hair and wearing brightly colored shirts and dresses. One was in mini denim coveralls. Their teachers directed them to sit in semicircular rows in the middle of the floor. I stood watching beside the slide screen. Their bright eyes checked me out. I smiled, "Hi."

Then came the kindergartners, led by their teacher, holding hands, all in a row. Smaller versions of the six-year-olds, they had chubby cheeks and an innocent glow.

They were positioned in front of the first graders in semi-circles two deep. Some sat cross-legged, others with knees clutched against their chests. *Oh, my. This will be an experience*, I realized.

I felt that familiar tinge of anxiety; the one when Mrs. Honess, my fourth grade teacher, called on me to give my first oral book report. I would really have to *think* about what I was saying—to choose very simple words. Might I bore these five- and six-year-olds? Would they soon be squirming, with anxious teachers imploring them to pay attention? *Yup, that's a distinct possibility*, I thought. I could bomb with a kindergarten class.

The teachers settled the chattering tykes and introduced me. I looked down on some 60 little faces all gawking straight back at me. I felt like Art Linkletter on his acclaimed TV program *House Party*, which aired from 1952 to 1970. It contained a segment called "Kids Say the Darndest Things" in which Linkletter interviewed schoolchildren between the ages of five and ten before a live audience. I recall him perched on a stool before several seated kids. Except for glances toward onlooking parents, they were trained on Art like cats on a caged canary. Their replies to his seemingly innocent questions were unpredictable and often hilarious. This was, of course, before the days of broadcast tape delays.

For example, Art asked one youngster if there was anything his mother had told him not to say on the show. He said his mom told him "not to let any skeletons out of the closet."

When Art asked for an example, he blurted, "My mom is going to have a baby but my father doesn't know."

I started tentatively, projecting a slide of a rabbit and then a deer onto the screen. I talked about where each lived and what it ate. The next slide was the menacing close-up I took of the black bear on the slopes of Bull Lake Canyon.

Big gasp by the kids. Several wriggled their bottoms farther from the slide screen.

"What is that?" I asked, deciding to let them tell me what *they* knew.

In unison, "A bear!"

"Where does it live?"

A cacophony of responses erupted. Among them I deciphered, "the woods," "the forest," "the mountains." Then, after all others were finished a small voice in the front trilled, "In my grampa's house." The four teachers and I laughed. The kids giggled and squealed. Grampa had a bear rug in his house. At least, that's what I was going to assume.

More slides flashed onto the screen—a beaver, an eagle, an elk. Then I showed a picture of a yellow-bellied marmot. A woodchuck relative, it's a six-pound rodent largely restricted to rocky hillsides in the Rocky Mountains. It was an animal most of these youngsters had likely never seen, even though they were not uncommon in the reservation's mountains.

By now the kids no longer waited for my question, "What's that?" They just shouted their answers as each new image materialized on the screen. Several times the teachers reminded them to raise their hands so the biologist could call on them. But some just couldn't contain their enthusiasm. So all followed suit, growing more boisterous as the hour progressed.

As the marmot flooded the screen, dozens of flapping tiny hands popped up. "Rabbit!" "Rat!" "Squirrel!" "Chipmunk!" And maybe some other creatures' names drowned out by the louder shouts.

"Children, children," a teacher pleaded. "One at a time, please!"

The mayhem subsided to silence. As I opened my mouth to speak, a thin voice announced, "That's Billy's dog."

The room erupted! I bent over, clutching the back of a folding chair. My stomach ached from laughter by now, but this was the topper. Kids were giggling, grabbing their neighbors. Inch by inch, they had all scooted forward on the floor, wriggling together into a knot. The closest—the kindergartners—almost touched my feet. Their round little faces with gap-toothed smiles and joyful eyes beamed up at me.

I managed to finish the slides, reinforcing the right answers after the kids had chorused their guesses. The "one at a time" rule never quite caught on.

"No, a skunk doesn't eat carrots," I responded. "It eats insects and mice mostly." Then I thought to myself, given a chance, I suppose a skunk would eat a carrot.

In all, the kids' knowledge about animals of the forests and mountains matched their enthusiasm. Or nearly so. A full hour sped by.

"Thank you all for being such good listeners. You made this a lot of fun." I realized I spoke as much for myself as for their experience.

I held up a book with bold pictures and stories about animals. "This is for you. I'll leave it with your teachers." Then as an afterthought I added, "I hope you will talk to your parents about the animals that live on your reservation. Thank you again."

"Thank you, Mr. Smith," they chimed, parroting a teacher.

Two teachers filed the chattering tots from the room in a bouncy row. The two others thanked me as we continued to laugh. "You know, the kids *really* enjoyed this," one darkhaired woman wearing a beaded Indian necklace said.

Then the one with no hint of native heritage jumped in, "I just hope you weren't overwhelmed."

"Well, maybe at first," I replied, "but then I let the kids show me how to talk to five- and six-year-olds."

I wondered what discussions around dining room tables that talk prompted. I would have paid money to eavesdrop. And moreover I wondered, *Did Billy's dog really look like a yellow-bellied marmot?* I would like to have seen that dog.

PART III

Treat the earth well: it was not given to you by your parents; it was loaned to you by your children.

—Ancient American Indian proverb

8

On the Same Page

I eagerly delivered reports—annual progress reports, habitat inventory summaries, status reports of big game species—to the Tribal Fish and Game Committee and Joint Business Council. Certain council members would thumb the pages and glance curiously at me as if to say, "I am not impressed by how much time you spend at a typewriter."

Most of these documents were skimmed at best. I knew that. But I regarded chronicling our findings an essential contribution for succeeding biologists, tribal leaders, and the Shoshone and Arapaho people. This permanent record was the benchmark by which future efforts to restore WRIR's natural treasures would be gauged.

I liked that my position provided access to tribal decision-makers. It is not that it conferred a sense of self-importance; rather it made my work feel relevant. Being at the Joint Business Council's beck and call and able to schedule a time slot to present some pressing matter of my work to this governing body was gratifying and often hastened decision making.

However, soon after arriving at the reservation I found myself in the uncomfortable position of asking the Joint Business Council for money. The Fish and Wildlife Service's annual commitment failed to match the size of the job at

WRIR. My salary, vehicle expenses, and a small operating budget fell far short of funding the sizeable inventory needs of a place the size of Yellowstone National Park—especially given the reservation's total lack of preexisting wildlife and habitat data.

But argue as I did for more operational funds, our supervisors in Billings, Montana, said no dice. "Do the best that you can. This is the budget you have to work with."

I was employed just two months when I first groveled. Dick and I requested funding from the Council to survey big game populations during the winter of 1978–79. I planned four replicate winter surveys of the Wind River Range and Owl Creek Mountains. Glancing up from the budget I passed out, former tribal game warden Alfred Ward asked "Why do you need to fly so much?"

A fair question, I thought. "I'd like to fly four times so that I can get an idea of how accurate a single survey might be, and in what month can I get the largest count. And with four surveys, I can map what areas elk and other game use throughout the winter. I can also measure what kind of calf losses ..."

"OK," Bob Harris interrupted. "But you also want $3,000 to count the antelope. That's a lot to count animals we don't have."

Besides the mountain flying, I had requested 30 hours of fixed-wing flights to survey pronghorns. The optimum time for counting pronghorns, mid-summer, was fast approaching.

"I know that's expensive. But the reservation has at least one and a half million acres of pronghorn range. Flying is the only sure way to learn how large or small the population really is."

I felt perspiration dampen my armpits. Damn! I just hated asking them for money. But even more, I didn't want

The author (in center pointing) on a field trip with members of the Joint Business Council in September 1978

to miss a whole year of survey work, knowing the Fish and Wildlife Service wasn't going to pony up.

There followed what seemed to me an interminably long silence. My stomach churned like it did that time when I was ten years old and rashly gobbled a pound and a half of fresh Bing cherries. Councilmen's eyes pored over the budget details I had typed. In hushed voices, some conferred with their neighbors. Then, as refusal seemed imminent, a potent voice came to the rescue.

Frank Enos and I had quickly become friends and confidants. No other individual made more important contributions to the cause of wildlife restoration in my years at WRIR. This was but one time when his influence moved conservation forward. "I think that we need to find out what we can about our big game herds. If we have to come up with the money for this year, then we should."

I could have kissed him.

After much recrimination of the Fish and Wildlife Service for not providing such funding, the Council agreed. The money was pledged, but with the proviso that this was a "one time" commitment for the upcoming year only.

My federal budget remained woefully inadequate throughout my tenure. Annual reports reminded our Billings office of that each year. And the job of inventorying wildlife across such vast lands required flight time, which was expensive then as it is now. In the end, the tribes unfailingly funded my flying *and* two seasonal technicians each successive year. The tribal councilmen, as a whole, were committed to restoring wildlife and showed it by obligating scarce tribal dollars.

🦌 🦌 🦌 🦌 🦌

I spent many solo mornings driving the reservation's paved, dirt, and two-track roads. In spring, I sometimes traveled to out-of-the-way stock ponds, denoted as blue smudges on the six-page map of the reservation's road system. Other puddles I happened to discover I added to the map. These forays generated lists of migrant and nesting shorebirds, wading birds, and waterfowl. Build a body of water and the birds will come. Serendipitously, I might spy an orange-crowned warbler flitting in nearby willows or a prairie falcon streaking overhead. Although tribal council members had expressed little concern about waterfowl or nongame birds, it was low-cost survey work that expanded our knowledge of WRIR's diversity. It also filled a gaping hole in Wyoming's Avian Atlas, a statewide catalogue of bird distribution. And, it was great fun.

In all, Kevin Berner and I surveyed 27 ponds and lakes exceeding 10 surface acres in size in the Wind River Basin, plus many smaller ones, and 68 miles of the main stem Wind River. These inventories led to recommendations on improving wetland and riparian habitats by maintaining water levels, fencing cattle away from pond-side vegetation, providing wildlife-friendly designs for new pond construction, and prohibiting hunting during critical nesting, fledging, and molting periods (March through August). In a status report on reservation waterfowl, Kevin and I cautioned against construction of two large impoundments.[1] Those proposals by private irrigation districts and the federal Bureau of Reclamation were destined to inundate hundreds of irreplaceable acres of wildlife-rich habitats. As water storage areas, each would be partially emptied during summer, thereby exposing a barren "bathtub ring." To date, tribal leaders have rebuffed such proposals that would exploit tribal waters.

During wetland surveys, I stumbled across places where effluents from oil wells were contaminating surface water and where birds lay caked in oil-slicked ponds. I retrieved a greased badger on one occasion, found a decaying muskrat on another. Tribal and BIA officials sought correction of both issues and got it in some cases. Wire lattice festooned with colored flagging was suspended over the most egregious lagoons to deter bird landings. Escapement of effluents became better controlled at other sites. These small remedial victories promised change in future behavior. They also fed my need for tangible progress in my partnership with the Wind River tribes.

🦌 🦌 🦌 🦌 🦌

Habitat mitigation and bird surveys are examples of how the job ripened as I worked toward preparing a reservation wildlife management plan. This was the "big picture" job I had dreamed of in college. The breadth and challenge of the work were limited only by the hours in a day. I wanted the management plan to be more than a recipe for restoring the reservation's big game. So I constantly explored ways to incorporate all wildlife. I'm forever grateful to the

Joint Business Council for granting me that freedom, and to Dick Baldes who indulged my persistent curiosity. Bald and golden eagles and a variety of hawks remained important in the culture and religion of both Arapaho and Shoshone people. Feathers, talons, and wing bones were all used in the annual Sun Dance, powwows, and other tribal ceremonies. It was unclear if the federal laws that protected these birds, such as the Migratory Bird Treaty and the Endangered Species Act, applied to tribal members on the reservation. The federal government contended yes. The WRIR, a semi-sovereign nation, asserted otherwise.

To supply the ceremonial needs of American Indians and to limit killing of birds, in 1972 the USFWS established a national repository of bird of prey remains—those killed by vehicles, electrocution, illegal shooting, or other means. Our office served as a liaison. We helped tribal members fill out the required paperwork so they could obtain raptor parts from the repository. This win-win approach provided cultural materials for Indian people while protecting birds of prey.

I visited the locations of historic bald eagle nests in cottonwood forests, but found no evidence of nesting. I did map active nests of golden eagles and various species of hawks. But the reservation was a big place, the size of Yellowstone National Park. To help locate nests of eagles and evidence of threatened and endangered species— including peregrine falcons, gray wolves, grizzly bears, and black-footed ferrets—I appealed to the public. I placed an article in the *Wind River Journal* asking readers to report sightings of these scarce animals to our office.

Just two reports of endangered species reached my desk—both of black-footed ferrets, the rarest mammal in North America. Both, unfortunately, proved to be cases of mistaken identity. However, the effort generated publicity for conservation. Mrs. Charles Snyder, a Crowheart resident and string reporter for Lander's *Wyoming State Journal* newspaper, wrote a piece for that paper titled, "Black-Footed Ferret Turns Out to Be Long-Tailed Weasel." She related my quest to learn of rare animals on the reservation. Then she recited her own effort, beckoning me to view several 35-millimeter slides her husband had taken of a slender animal scampering near their house. No ferret as it turned out, but good PR and great cookies.

🦌 🦌 🦌 🦌 🦌

Opportunities to help wildlife were often serendipitous. After a year on the job, reservation residents increasingly contacted our office with questions and concerns about wildlife. It became a kind of animal psychic hotline. I saw each contact as an opportunity for two-way learning. Positive experiences would hopefully be shared with each caller's family and neighbors, growing trust in our efforts to aid reservation resources. I, in turn, gained new insights into the people's interests and affection for wildlife, and encountered some kindly people I would otherwise not have met.

Not all contacts were friendly, however. One tribal member, Tommy LeBeau, accused me in a phone call of harassing wildlife with aircraft. I listened, then began explaining the purpose of our surveys and the precautions we took. Before I got far, he warned me, "You don't belong on the reservation." My next encounter with Tommy several months later would be face to face.

Rescue calls were common in spring. Baby robins that prematurely tumbled from nests and birds crippled or knocked dizzy by run-ins with windows topped the list. These came from Indians and non-Indians alike, although the local State Game and Fish Department's office fielded the lion's share of calls from Lander residents. Generally I could do little for the animals in question. But there were exceptions.

A male American kestrel

One incident stands out after all these years. A Shoshone woman called one August day and told me a hawk had crashed into her house. It lay dazed in the yard. Hawks rarely fly into windows, so my curiosity was piqued.

"Are you sure it's a hawk?" I asked.
"Yes, I think so," she said.
"Do you know what kind of hawk?"
"No." She didn't know what it was called.
"Is it big or small? For example, compared to a robin."
"Maybe robin-sized."
"What does it look like?" I continued.
"Very pretty, brown and white and black and blue," she replied. She repeated that she was afraid something would kill it (probably her dog, I guessed) in its vulnerable state.
"Blue where, on the wings?" I prodded.
"Yes, on the wings."

Oh yeah, I could break away from my office work for an hour. Most rescue calls led me to dumbstruck sparrows, pine siskins, finches, and flickers. This promised to be no misguided robin seeking worms in a potted palm.

Ray Nation and I drove to the location she described outside Fort Washakie. Below a picture window quivered a kestrel, widely known in North America as a sparrow hawk. It was a male, unmistakably defined by its rufous back and tail and slate blue wing coverts. This bird was as strikingly colored and marked, with black facial streaks and bold belly spots, as it was extraordinarily engineered for its occupation as predator.

I was once asked, if I were to be reincarnated as a bird, what species I would choose. The answer was easy—a kestrel. Beyond the brash looks, kestrels are acrobatic aviators. This smallest of North American falcons is both crafty and versatile in pursuit of its prey—principally mice, voles, large insects, and small birds. It hunts from perches or while hovering, or by foraging on foot. Cruising 20 feet above a field, a kestrel screeches to a hover, then plunges deftly on a hapless grasshopper. A daily diet of two plump voles or two handfuls of hoppers makes kestrels the farmer's altruistic ally.

With wings held wide and beak agape, the kestrel on the woman's lawn put up a fierce front. His defiant dark eyes darted from Ray to me, and his head robotically tracked as I circled opposite Ray. But whether stunned, confused, or frozen by fright, he didn't retreat as we closed in. It proved easy and painless to drape him in burlap, but less so to untangle his talons once in the cardboard box.

We assured the relieved woman that we would try to nurse the bird to health. And yes, I would let her know the outcome. Guessing this might end badly for the falcon, calling wasn't something I would have offered. I would do it only because she asked.

So off Ray and I drove, wondering what we had gotten ourselves into. Neither of us had experience rehabilitating an injured bird. But I was optimistic, if only because neither the kestrel's wings nor legs looked or felt broken. If they had been, recovery would have required a veterinarian's attention and a burdensome recuperation, with no certainty that the bird could ever succeed again in the wild.

Back in Lander, we showed our charge to an office of skeptics. Undeterred, we placed a bowl of water in the box and dashed into a nearby field. I knew the bird had been in the woman's yard overnight, likely without food. Stuffing a jar with brown- and yellow-winged grasshoppers, we hoped to remedy that.

The hoppers proved a hit with the little kestrel. Ray popped open a flap of the box and I tossed one in. The first one propelled itself straight back past our faces. The next one we disabled to give the kestrel a fair chance. Snap! Talons shot forward engulfing the prey. Then the hooked beak went to work and off with its head! In 30 seconds our charge had dispatched the first course. Another hopper into the box. Another hopper down the hatch. The little kestrel cocked his head and peered up, awaiting course number three. This was fascinating to watch from just two feet away. Balanced on one leg—an encouraging sign of his motor function—he lifted each clutched insect to his hooked beak, then devoured it in several bites. Ray and I grinned like grade school kids feeding worms to a nest of downy robin chicks.

I'm not certain how much a kestrel eats per day—maybe a quarter of its four-ounce body weight, maybe more. This one ate a dozen inch-and-a-half-long grasshoppers in less than an hour. We ended the feast for fear of overfeeding him. But I was elated that perhaps the biggest obstacle that we might face in his recovery was no obstacle at all. If he didn't OD on grasshoppers, food would be no problem.

When I arrived at the office the next two mornings, I immediately cracked open a flap of the box perched on a table beside my desk. There he was, alert and brashly presenting his wicked, hooked beak. I filled the bowl from a watering can to limit my exposure and his excitement over a hand violating his space. Then it was out to the field for another course of hopper helper.

When I entered my office the third morning, the little sparrow hawk was beating his wings inside his cardboard cell. He was telling me, I surmised, that he had tired of room service. Time had come to catch breakfast on his own. I peeked in. An attempted escape confirmed his alacrity.

Ray and I carried the box to the field, with our colleagues as spectators. On opposite sides, we opened the box flaps. The kestrel looked up at us with wide, fiery eyes. As I put a gloved hand inside to lift him out, he clamped a finger in his beak. Good sign, I thought. Rather than lift him out and risk retaliation for his imprisonment, I knelt on the ground and rolled the box on its side. The kestrel hopped out and flapped into the morning sun, whisked on his way by our wild applause.

"What do you think, a career change to professional bird rehabilitator?" Ray laughed as the kestrel's blue wings dissolved into the sky.

"Oh no, not for me. I'll stick with what I know," I replied, noting the beak imprint in my glove.

Ray and I couldn't help feeling self-congratulatory. Saving one bird was a modest triumph, but one that would warm another's heart. While the others lingered in the morning glow, I returned to my desk and called the Shoshone woman to share the good news.

🦌 🦌 🦌 🦌 🦌

I mentioned an unpleasant phone conversation I had with a Shoshone named Tommy LeBeau. His intimidation went no further, although I didn't take it lightly. Instead, he helped remedy a catastrophe that struck WRIR's vulnerable pronghorns.

Tommy and his Arapaho friend Jimmy Coulston were big-time hunters on the reservation, and two big reasons that hunting restrictions were needed. While writing this chapter, I e-mailed my former field technician, Kevin Berner, to confirm some details about his experiences. He answered my request and added a story that he considered the low point of his time at WRIR.

Fellow field technicians Ray Nation and Philip Mesteth, both American Indians, joked about fair-skinned Kevin meeting Coulston someday. "Coulston's a bad actor. He has a reputation for hating non-Indians," Ray told him.

As Kevin tells it, "One day, Ray, Phil, and I were driving into the Wind River Range to do some vegetation sampling. I was seated in the middle on the pickup's bench seat."

Whether Kevin knew it or not, real cowboys consider this the choice seating position. Riding in the middle, you don't have to drive, and you don't have to get out to open and close gates. It also proved beneficial to Kevin's welfare.

Anyway, Kevin was in the middle by choice or default. "Suddenly, another truck approached and forced our Fish and Wildlife Service truck off the road. The driver, a big Indian, leaped out and charged over to us. He was pointing and shouting at Phil and Ray, 'Get that white nigger off my land!'"

At that moment, Kevin still didn't know who this wild man was, but he was quite sure the two of them would never be close friends. After completing his tirade, Jimmy stormed off.

"Who was that?" a wide-eyed Kevin asked.

"You just met Jimmy Coulston," Phil laughed.

Kevin reflected, "I felt lucky to have been with Ray and Phil that day. I heard other stories, some from Dick, that made Coulston sound less than stable. Later, when Dick learned of our encounter, he insisted we notify the FBI because Coulston was 'impeding federal government employees.'"

So off Dick and Kevin went to Fort Washakie to talk to the man. That, too, Kevin found unsettling, knowing that the next time he bumped into him, Coulston might be further motivated to bodily remove Kevin from the reservation. Fortunately for Kevin, there was no "next time."

🦌 🦌 🦌 🦌 🦌

According to the National Weather Service in Lander, the winter of 1978–79 was the coldest on record. Monthly November, December, and January temperatures averaged 11 to 18 degrees colder than normal, and February was five degrees colder than normal. Moreover, snowfall was the second largest in the 95 years of record keeping. The severe conditions jeopardized survival of livestock and wildlife across Wyoming.

Pronghorn herds south and east of WRIR were particularly hard hit. During late November and December, over

Pronghorn antelope tangled in fence along the Sand Draw Road, March 1979

1,000 pronghorns migrated into the southeast corner of the reservation. Snow buried the vegetation there, including most sagebrush, from mid-November through January. In early December, I received reports of pronghorns in trouble. By mid-December, the local state game warden, Tim Britt, and I saw a disaster in the making. En route to the protection of cottonwood forests and draws along the Little Wind River—atypical pronghorn habitats—dozens of pronghorns met gruesome fates.

Four to five hundred had balled up near the junction of the Sand Draw and Gas Hills highways. That intersection proved a grisly death trap. Pronghorns negotiate four- and five-strand barbed wire fences—the kind so common across western rangelands—by crawling underneath. They rarely jump these fences and are reluctant to squeeze between tightly spaced strands. When they do, they often lose quantities of hair, receive wire cuts, or become hopelessly tangled—effortless food for coyotes, eagles, and ravens.

The scene that Britt and I found was utterly sickening. Snow had buried the highways' bottom fence wire, then had crusted. Pronghorns trailed along the fences to the fateful highway intersection. We found animals entangled and dozens more frozen stiff within yards of the fences. Still others lay dead along the roadsides; they had made it through one roadside fence, but, unable to negotiate the opposite right-of-way fence, they had panicked and been battered by vehicles on icy roads.

I called the Wyoming Department of Transportation for permission to modify the right-of-way fences. On December 14, my office staff and I went to work. We pulled staples and laid down fence strands at six locations along a six-mile stretch of the Sand Draw Highway. This allowed hundreds of weakened pronghorns to pass through. Still, I found drained and exhausted animals perishing as the last snow melted in April.

On January 8, 1979, at the depths of the disaster, I made a reconnaissance flight over the area. Fifty-nine percent of the 246 pronghorns I observed were in deep draws where big sagebrush and greasewood, shrubs that reached heights of three to six feet, were the only food above the snow for miles. The land looked arctic.

Adjacent to the reservation, conditions were no better. By winter's end Tim had investigated 249 carcasses, half of them fawns. He believed that hypothermia was the primary killer when temperatures commonly plunged below -30 degrees during December and January. In subsequent months, hunger took over. On March 22, I cracked opened the femurs of nine frozen carcasses. The marrow inside was red and gelatinous, the stored and normally white fat totally depleted from malnutrition.

The extent of the losses wasn't fully apparent until August 1979. The previous summer I surveyed all 2,500 square miles of potential pronghorn habitat on WRIR. Over several mornings in late July and early August 1978, a second observer and I flew 30 hours in Larry Hastings' Cessna 182. Each morning we flew mile-wide north-south transects, progressing across the reservation until all suitable habitat was covered. The same rigorous protocols were repeated on surveys each year: same period of time, same transects, same aircraft, pilot, and observers. When I compared results from 1979 to 1978—after subtracting the new fawns observed from the 1979 total—the reservation's pronghorns had declined by 54 percent. Not only had half the population died, but the gestating fawn crop suffered as well—a delayed effect of the killer winter. Reproduction, as measured by the number of fawns observed per 100 does, was 30 percent lower in 1979 than in 1978. Pronghorns on lands adjacent to WRIR suffered even greater losses: 60 to 80 percent, according to Britt.[2]

These were epic conditions and of course nothing could prevent future severe winters. But those deadly fences were another story. That ghastly ordeal just couldn't be repeated.

I first prepared a detailed report on the winter's toll, then did media interviews, conferred with Tribal Fish and Game Committee members, and asked to present a recommendation to the Joint Business Council. On November 8, 1979, the Council heard my impassioned request. I prescribed raising the bottom strand of barbed wire fences from the existing 10 inches to 16 inches above the ground wherever pronghorn migrations crossed the Gas Hills and Sand Draw highways. This fence design—recommended by the BLM for the West's public rangelands—contained livestock, but enabled pronghorns to crawl underneath. In a unanimous vote, the Council approved the change and assigned me a work crew of four employees from CETA (Comprehensive

Employment and Training Act, a federal jobs training program). With the approval of the Wyoming Department of Transportation, we began work on November 19. In ten days, we changed nine miles of fence, facilitating the free movement of pronghorns where bleached skeletons memorialized last winter's tragedy.

Tommy LeBeau was one of the CETA workers. We had spoken just that one time on the phone, but he made sure on the first day of fence work that I knew who he was. As we worked side by side—yanking staples with fencing pliers, then reattaching the raised bottom strand—Tommy regaled me with hunting stories, including the day he and Jimmy Colston killed 17 elk in Bull Lake Canyon.

"All the meat went to people who needed it," he declared with a shot from cold, dusky eyes. "We don't need your damned game code."

I let his words settle, then replied, "I'm glad the meat got used. It's too valuable to waste."

Looking from the pliers he clutched to the clenched jaw that had failed to smile all morning, I added, "Let me tell you one thing I believe. A reservation game code should allow able hunters to fill the hunting permits of members unable to hunt for themselves. That tradition of your people is important to maintain."

He looked curiously at me for a long moment, perhaps surprised at the words and the sincerity he found in my face. Without a word, he raised the pliers as if in slow motion—then popped another staple from a post.

I don't know if Tommy was among those Shoshones who voted for the game code the following fall. Probably not. Regardless, he had lent a hand to the reservation's pronghorns. Perhaps our collaboration at least kept him from turning others against the progress my office and tribal officials were making. I like to think so.

9

Game Code

The biggest challenge to restoring and sustaining wildlife on WRIR was gaining control of tribal hunting. The 1948 game code, the only previous comprehensive curb on hunting, lasted just five years. The accounts of elders painted a bleak picture of wildlife decline following its repeal in 1953.

Fearing that further depletion of game was likely, the Shoshone General Council appointed a committee of men and women to draft a new game code in June 1978. This was presented by Chairman Frank Enos in February 1979 to the Tribal Fish and Game Committee, composed of three Shoshone and three Arapaho councilmen. The councilmen enlisted my help to review its contents. We redrafted the code several times, largely to mediate differences between the two tribes' representatives. Provisions governing game birds and waterfowl were dropped. Protection of federally threatened and endangered species was avoided. The final document addressed the hunting of only nonpredatory big game species: mule deer, white-tailed deer, elk, moose, bighorn sheep, and pronghorns. During the fall of 1980, the six-member Shoshone Tribal Council approved the game code and recommended its presentation at the Shoshone General Council meeting on November 8. Dissension persisted from some Arapaho councilmen. Tension between certain members of the two tribal councils was palpable. One bluntly insisted, "Arapahos don't need any white man's laws."

Prior to the Shoshone General Council meeting, I wrote an article for the *Wind River Journal* about the decline of game on the reservation. I described how the tribes could reverse that trend. Next to my article, the "Tribal Viewpoint" asked, "Do you think there should be some sort of tribal game code to preserve our wildlife?" This is one of those "man on the street" weekly questions that many newspapers randomly pose to people encountered in the post office or cornered at local businesses. All four Indians who responded agreed that such regulations were needed. John Gavin, a former Shoshone game warden, answered, "All hunting should be shut down for at least five years so that the deer herd can be built up and also the moose."[1]

The Shoshone councilmen felt confident the game code would pass. Frank Enos presented the proposal at the well-attended November 8 meeting. Dick told me later that Dr. Enos did a superb job. Limited discussion ensued. In a decisive declaration, the General Council approved the game code by a margin of two to one. The vote was a due victory for all who worked so diligently for its adoption, and represented a potential turning point for the WRIR's beleaguered wildlife.

I was ecstatic to hear the news the following day. Fieldwork is what we biologists are trained for and love to do. But less than half a biologist's work hours are spent communing with nature. Advancing the cause of conservation is the payoff for long hours of data analyses, literature research, writing, citizen and media communications, and taxing public meetings. I couldn't stop smiling for days.

The game code established a hunting season from September 15 to December 31; a limit of one big game animal of each species per enrolled member; a free licensing system; a tagging procedure requiring licensed hunters to obtain game tags before hunting and to tag animals immediately after harvest; the prohibition of snowmobiles, motorcycles, and aircraft for hunting; and a 10-day, preceremony hunting season of antlered deer and elk for participants in the Sun Dance (the Shoshones' most important religious ceremony, observed near the summer solstice). Big game of either sex could be killed during the fall season. Game tags were transferable from one licensed hunter to another, so Tommy LeBeau and others could hunt for their relatives. No restrictions governed species other than big game.

But wait—time out for a reality check. *Both* tribes had to pass any new proposal for it to become the law of the land. The likelihood of the Arapaho General Council adopting the game code was shaky. Arapaho councilmen were getting plenty of feedback, as a growing undercurrent of hostility suggested tribal members were less than enthusiastic about restrictions on hunting. Dick and I couldn't gauge if this was a vocal minority or a more pervasive attitude.

My fledgling outreach to schoolchildren might yield future benefits for reservation wildlife. But only tribal members 14 and older could vote in the upcoming General Council meeting. Before the spring meeting, I wrote a second article for the *Wind River Journal* urging Arapahos to place limits on tribal hunting to restore herds and improve future hunting prospects. Now I wondered if a more personal appeal to adults might prove helpful; maybe a public forum, a way to allay people's concerns about the intended purposes of a game code. But what were the downsides?

I asked Arapaho councilmen Emil O'Neal and Ernest SunRhodes, "What do you think of an evening informational meeting? Would that be a good way to reach the Arapaho people?"

Neither rejected the idea offhand. Both were skeptical that anyone would attend. Ernest joked, "You better advertise plenty of food and drinks."

"On the other hand," Emil teased, "you'll probably survive the evening."

Seemed like a rousing endorsement.

Actually, O'Neal and SunRhodes offered to arrange a venue and advertise the talk. "As long as we're doing it, let's plan on two evenings!" Emil added.

So I drafted the following announcement that the Arapaho Council placed in the *Wind River Journal*:

> A talk and slide show concerning ongoing studies of the wildlife resources of Wind River Indian Reservation will be presented this week. Information on the size and trend of big game populations, migration patterns, and management needs will be covered. A discussion will follow.
>
> The Great Plains Hall at Arapaho, Wyoming, will be the location of this presentation on Thursday, May 14, from 8–9 p.m. The same talk will be given at the Blue Sky Hall in Ethete on Friday, May 15, from 8–9 p.m. The Arthur-Antelope-Brown American Legion Post #84 of Arapaho is sponsoring this community service. REFRESHMENTS WILL BE SERVED. Anyone who is interested in the future of the reservation's wildlife is encouraged to attend.

As we set up a slide projector and screen in Great Plains Hall, a tug-of-war of conflicting scenarios preoccupied me. "What kind of turnout do you think we'll get?"

"Hard to say. But I think we've got plenty of food," Dick laughed.

"Hey, let's put the refreshment table just inside the front door," I suggested. "We can use it to chum any curious onlookers inside."

That first evening met the low end of our expectations. I preached mostly to the choir. Not counting Emil, Ernest, Dick, and myself, we had three in attendance. One of those was a staff writer for the *Riverton Ranger*, a newspaper that carried reservation news. Dan Neal was kind enough to write a half-page article in Friday's paper, complete with photos and interviews of Emil and Ernest. Quotes from both Arapaho leaders urged adopting the game code.[2]

"If we don't protect them [the reservation game herds] we're not going to have anything in ten to twenty years," claimed Councilman O'Neal. "We're not going to have anything in ten to twenty years, just like the buffalo," he said. "The only thing increasing on the reservation is prairie dogs."

Neal wrote that 74 percent of Arapahos responding to a wildlife survey said they believed that reservation herds had declined in recent years. "If you really feel the populations are down, then there are a lot of things we can start working on to get those populations back up," I was quoted as saying near the end of my presentation.

The survey that Dan Neal referred to—the one that the Tribal Fish and Game Committee and I devised—provided persuasive evidence that members of both tribes recognized the plight of wildlife. This was a clarion call to action; an impetus to improve conditions for the current and future generations. But distrust is a self-serving ruler. Shoshone passage of the game code disaffected Arapahos and jeopardized its acceptance. Moreover, most tribal members were too young to have lived under WRIR's 1948–1953 game code. They were unconvinced of the worth of hunting constraints. As Samuel Johnson noted in 1755, "Change is not made without inconvenience, even from worse to better."

Neal's article announced the second informational meeting at the Blue Sky Hall in Ethete the following night. "Smoked fish and coffee will be served," the piece announced. In Ethete, there was also plenty of food and drink for another audience of just three. At least they were three different people.

🦌 🦌 🦌 🦌 🦌

The Arapaho General Council was scheduled to consider the game code by the end of 1980. After several delays, the hearing and vote were rescheduled for June 13, 1981, at Blue Sky Hall. Auspiciously, that wasn't a Friday. Under consideration the same night was repeal of a controversial issue regarding tribal enrollment of children. It was an unfortunate circumstance. Enrollment, always a contentious topic, might poison the game code's adoption. The tension was so intense that one member remarked, "The Arapaho wouldn't have passed anything that night."

I remember like yesterday entering Blue Sky Hall that night. The auditorium's bleachers were filled with more Indians than I had ever seen at one time—even more than attended a Wyoming Indian High School basketball game. And that was a state championship team! Opposite the bleachers was a stage crowned in linen-covered tables placed end to end. A boisterous din echoed from the cement block walls. The atmosphere was electric.

Dick and I were asked to attend the meeting; Dick, to help answer questions; me, to give a slide presentation about how our wildlife inventories provided justification for hunting regulations. That was to precede a presentation of the proposed game code. Arapaho councilmen would cover that, and a vote on adoption would follow.

The game code had been scheduled first on the agenda, but by popular demand was bumped to second. The new first item—the enrollment matter—got the meeting off to a bang with emotional testimony interrupted by audience outbursts, pro and con. My stomach began to churn. What was I doing here? Why had I offered to make a presentation?

Why did the Arapaho Council accept? This audience was in no mood for a dog and pony show on how to manage their reservation. This would be like teaching a cat to sing—a waste of my time and deeply annoying to the cat.

After the enrollment provision was soundly revoked, Councilman O'Neal rose to introduce the next item. In barely two minutes, he said that the game code passed by the Shoshones (loud jeers from audience) was up for a vote next. "But first, Bruce Smith, the Fish and Wildlife Service biologist, will talk about game populations and then go over the game code."

Wait! I silently shouted. *I'm not supposed to cover the game code, just give a slide talk on our station's work.*

The lights were quickly dimmed, yet I could see young men in the first couple of rows chattering and gesturing toward me. I felt at once that I was the least popular white man in the auditorium; the only others present being a bunch of Denver lawyers safely removed atop the stage. They represented the tribes in disputes over water rights and the pilfering of tribal oil and gas royalties by unscrupulous corporations. From my desolate folding chair at the base of the stage, I could see them flanked by the Arapaho councilmen. In starchy white shirts, stylish suits, and striped ties, they tended notepads and briefcases. I, on the other hand, wore my best pair of Levi's and a plaid wool shirt and clutched the slide projector's remote. On reflection, I envision the *Car Talk* radio program's law firm, Dewey, Cheetham, and Howe. Yet I felt sure the lawyers were the crowd favorite compared to me.

One hard swallow, and I heard my voice: "The Wind River is blessed with lush habitats, but is starved for animals." The talk ran 15 minutes. It should have been five. Finally, the last slide flashed onto the screen—a solitary, backlit elk on a remote, snowy ridge. I felt my throat tighten. *This is no time to falter; the game code's next.* I glanced at Emil, firmly planted in his chair. No relief to be found there. So I summoned my trusty barometer of adversity. The Marines had shaved my head and sent me to war 12 years before. I survived that. This, I told myself, paled by comparison—although the odds looked worse today.

Humility and skepticism rode opposite shoulders as I asked for the lights. Now, this non-Indian would sketch how the Shoshone game code would restrict the Arapaho people's vested hunting rights, rights guaranteed by long-standing treaty. I hastily ticked off the code's key provisions. Then I asked those gathered to adopt it for the sake of future generations of their people. The catcalls that resounded came as no surprise.

I passed the microphone to Mr. O'Neal and retreated like a suspect from the witness box. "Now the vote will be taken," Emil announced as the clamor crescendoed. "All those in favor of the game code, stand to be counted."

No secret ballot here. Members of the Arapaho Entertainment Committee (how fitting, I thought) counted those standing in each section of bleachers. I could see Jimmy Coulston (Kevin Berner's buddy, you'll recall) and several others exhorting the crowd to stay seated, and I searingly knew why tribal councilmen were reluctant to take all but obligatory issues to the General Council.

Of 725 registered voters in attendance, fewer than a third rose to their feet. Game code adoption was crushed, as were my spirits. I recalled reading that when Winston Churchill lost his bid for reelection as prime minister, his wife consoled him by suggesting the defeat was a blessing in disguise.

"If so," Churchill responded, "then it is very effectively disguised."

Dick and I left immediately. Cheers following the announcement of the motion's defeat chased us out the door. I kept replaying the scene during the drive back to

Lander. Did I overcook the slide talk? What should I have done differently? As a non-Indian advocating that tribal members curtail their treaty rights, *I* would have resented me if *I* were Arapaho.

In a subsequent article, the environmental newspaper *High Country News* reviewed wildlife's troubles on WRIR. Emil O'Neal, a staunch supporter but vain spokesman for the game code, summed up its downfall: "The Arapahos voted against it because they thought it would greatly infringe on their rights."[3]

Pius Moss, the Arapaho elder I had interviewed months earlier, bemoaned the game code's rejection. "We need conservation so that we can save wildlife for the young. You have to go to the edges of the reservation now to find wildlife."

In the same article, Arapaho Jimmy Coulston said he asked the General Council to reject the code because he didn't want Baldes, Smith, or anyone from the state interfering in reservation game problems. Both he and his friend LeBeau did not believe there was any danger of eliminating reservation wildlife. "We are the ones who know where and how much game is on the reservation."

Throughout summer I came to accept the inevitability of General Council decision. The time, place, and conditions weren't right. Too many were not yet ready to hear and accept what must be done. I let it go, or tried to.

🦌 🦌 🦌 🦌 🦌

Following the game code's defeat, the Arapaho Business Council assured us that the matter would be raised again by year's end. Despite Dick's lobbying behind the scenes, that failed to happen.

I immersed myself in my field studies and writing. Time was short. I learned in late 1981, not long after returning from my father's funeral in Michigan, that funding for my position on the WRIR would end next June. Between November 1981 and June 1982, I cranked out a series of lengthy reports on the history, current status, and needed management of reservation wildlife: pronghorns, moose, bighorn, white-tailed and mule deer, and waterfowl. They modeled the account I had already written about elk. It kept my mind busy and distracted me from preoccupation and a lingering remorse that I had spent too little time with my father during his two-and-a-half-year battle for life.

I plugged away feverishly most evenings and every weekend that spring to complete this written documentation of my work. As May approached, I felt the crush of finality. Sand was slipping through the hourglass. Completing a reservation management plan by early June—my foremost assignment—was out of the question.

I had envisioned a final document streamlined to 100 or fewer pages with supporting information packed in the six wildlife status reports. Still, the plan would be a comprehensive blueprint for all major groups of wildlife and their habitats—no trivial task to be sure.

I petitioned Dick and our Billings office for more time. "Arrangements are finalized," I was told. "You report to your new duty station the second week of June."

I heard the words, but was hardly ready to leave.

Two months after the Wind River country faded from my rearview mirror, I completed the reservation management plan. The Lander office arranged its printing and distribution in September 1982. Though immensely pleased with the final product, I was disappointed that I couldn't assist the tribes with its implementation. During my last months on the reservation, I sensed growing sympathy for limits on hunting. I wanted one more run at game code passage.

The defeat at Blue Sky Hall had lost its sting and failed to taint my remarkable four years among the Arapaho and Shoshone people. The WRIR was an unrivaled opportunity.

During a 30-year federal wildlife career, it remains the assignment that gave me the greatest satisfaction.

🦌 🦌 🦌 🦌 🦌

Over the next two and a half years, wildlife management came to WRIR, albeit begrudgingly. Shortly before my departure, the Shoshone Council petitioned Interior Secretary James Watt to place a hunting moratorium on the reservation until the Arapaho General Council adopted hunting restrictions. Watt refused.

The winter of 1983–84 began much like winter 1978–79. Heavy snows in early December drove 2,000 elk and deer out of the Wind River Range to the foothills. Dozens became easy targets. Gary Lajeunesse and Dick Baldes reported in local news media that they knew of 139 killed during the first 10 days of December. But the number alone was not the worst of it. During a three-day period, 56 elk were killed within view of passersby on U.S. Highway 287; some were picked off from snowmobiles.

The story spread like wildfire. Headlines in regional newspapers read "Elk Slaughter Jars the Senses," "Elk on Reservation Killed by Poachers," and "Big Game Slaughter Sparks Tribal Discussion." Accounts in newspapers as far-removed as *The New York Times* reported the gory details, sometimes embellished by inflated body counts.

In a joint press release issued December 9, Business Council co-Chairmen Bob Harris and Wayne Felter admonished, "We must respect our wild game as our forefathers and elders have taught us." They added that the single hunter who killed fourteen elk and eight deer and others like him "seem to have lost their respect for our big game herds and their importance to us as Indian people for now and the future."[4]

That same day in a hastily scheduled meeting, the Joint Business Council discussed imposing an emergency moratorium on hunting. In the closed-door session that barred even enrolled tribal members from attending, no agreement was reached on the unprecedented proposal. The opponents rejected the measure as an infringement on sovereign rights, the same fatal drumbeat that echoed through Blue Sky Hall. Meanwhile, Arapaho and Shoshone elders made emotional pleas in the *Riverton Ranger* for swift adoption of a game code to protect tribal wildlife. In a foretelling of future events, one urged, "Something has to be done today, not tomorrow. Pretty soon we're not going to have any [wildlife] and the government will step in if the tribes don't settle it."[5] Days later, in the December 20 edition of *Wind River Journal,* a guest editorial appeared that was signed by a group of tribal members. They announced they were circulating a petition for adoption of a game code.

In a newspaper interview two months earlier, Dick had appealed to the Arapaho Tribe to adopt the game code. Reciting forecasts in the reservation management plan, he had stated that a game code could "triple the deer, moose, bighorn sheep, and antelope populations and we could easily double the elk."[6]

Contrary to much of Wyoming and the West, habitat loss and fragmentation were minor threats to wildlife on WRIR. The tribes had previously taken prescient steps to protect reservation lands. They established the Wind River Roadless Area in 1938, protecting from despoliation some of the most spectacular lands in the lower 48 states. They placed a moratorium on logging of reservation lands in 1971 to curb deforestation and road construction. And they restricted oil and gas exploration and development to certain arid lands of the Wind River Basin. These and other tribal actions favored their heritage over economic expedience and exemplified the land ethic Aldo Leopold advocated in *A Sand County Almanac.*

> The 'key-log' which must be moved to release the evolutionary process for a land ethic is simply this: quit thinking about decent land use as solely an economic problem. Examine each question in terms of what is ethically and esthetically right, as well as what is economically expedient. A thing is right when it tends to preserve the integrity, stability, and beauty of the biotic community. It is wrong when it tends otherwise.[7]

Whether by deliberate calculation or not, generations of Shoshones and Arapahos stewarded their homelands long before Aldo Leopold articulated his call for a communal land ethic. Their nomadic lifestyles, primitive weapons, and constrained populations accommodated bountiful wildlife. Just as surely, American Indians honored Mother Earth's gifts that the Creator provided for their use. Now in the late twentieth century, the tribes needed to reconcile their vanishing heritage with their tenaciously held hunting rights to secure wildlife's future.

Within days of the December 1983 elk slayings, the Shoshone Council took action. In a letter to Wyoming's U.S. Congressman Dick Cheney (future Vice President Cheney), the Shoshones requested his assistance. The Council urged him to encourage Secretary of the Interior William Clark (James Watt having recently resigned his post) to place an immediate moratorium on hunting of reservation game herds. As justification for federal assistance, the letter referenced the recent slaughter of reservation elk and deer, and invoked the failure of the Arapaho Tribe to adopt the game code.

When contacted by the *Casper Star Tribune*, Arapaho Council Chairman Felter said he was not aware of a request for a moratorium. Despite cosigning the press release a week earlier with Shoshone Chairman Bob Harris, he said, "I have yet to see any evidence of any slaughter." He added that he had not read the statement to the press that he signed.[8]

Felter told the *Star Tribune* that game conservation was the objective of both tribes, yet the Arapaho General Council had twice turned down a game code. That first occurred at the June 13, 1981, meeting (where I had been so persuasive). The second was a simpler code (no bag limits or restrictions on sex of animals) drafted by tribal lawyers that was tabled by the Council. Felter predicted there would be no federal moratorium on hunting, because no legal authority existed for such action.[8]

Before resigning, Secretary Watt reversed his earlier decision by ordering Interior officials to formulate and impose a federal game code on WRIR. He made clear it was to be the "toughest game code in the country," according to Fish and Wildlife Service personnel I interviewed. Then an editorial by Lonnie Williamson appeared in a 1984 issue of *Outdoor Life Magazine*.[9] It was titled "No Laws for the Indians." Williamson's piece drew national attention to what he called "the bloodbath" on the WRIR. His concluding remark that Native American treaties should be amended "so that all Americans must abide by state and federal wildlife conservation laws" was poorly received, especially by the Arapaho Tribe. That might be an understatement. It inflamed the ongoing dispute over game management between the two tribes.

Yes, things were heating up as I monitored the news reports with rapt curiosity. But the mercury would rise still higher.

The Shoshone Council persisted in its pursuit of game protection through BIA officials and at higher levels within the Interior. Finally, in July 1984, the BIA announced that it would impose a federal game code on WRIR. Not only that, the Office of Management and Budget authorized $140,000 to enforce it with Indian game wardens *from other reservations.* Topping this off, the federal game code was far more restrictive than the one the Shoshone Tribe adopted. It authorized just an October and November hunting season (I suspect the previous fall's hunting debacle eliminated December); allowed tribal members to take only one elk, deer, and pronghorn; prohibited hunting of all other big game; and placed restrictions on hunting of game birds and waterfowl.

This closely resembled the original concept I fleshed out with the Tribal Fish and Game Committee. It was considered unacceptably restrictive by leaders of both tribes in 1978. Six years later, Shoshone Chairman Bob Harris remarked that although most Shoshones were not in complete agreement with the regulations, they were pleased that a game code was finally in place.

The same could not be said of the Arapaho Tribe. For example, Arapaho Chester Aramajo, Jr., an American Indian Movement leader, declared, "We will hunt anyway. And if arrested, that will add fuel to the fire."[10]

Before the federal game code took effect on October 5, 1984, the Arapaho General Council again voted on and again rejected the Shoshone-adopted version, albeit by a much narrower vote than in 1981. Now the legal games began.

First the Arapaho attorneys hired an independent wildlife biologist to review all pertinent information the USFWS possessed. They claimed that there was "insufficient evidence confirming the declining numbers [of wildlife]" to warrant hunting restrictions. *What a waste of Arapaho tribal funds,* I thought. Our data were convincing, above reproach. They filled hundreds of pages in a dozen reports long since provided to the Joint Business Council. Most dissident tribal members did not refute that game populations were declining; rather they only opposed a perceived abdication of their vested rights.

Nonetheless, the Arapaho tribal attorneys filed a motion with U.S. District Judge Ewing Kerr. They sought a restraining order to prevent enforcement of the game code. The filing noted that this was the first instance of the BIA imposing a wildlife code on an Indian reservation. In oral arguments heard in Cheyenne on October 24 and 25, an Arapaho attorney challenged the authority of the Interior Department to impose restrictions on tribal treaty rights, arguing that a code was unnecessary and would harm Arapahos who wouldn't be able to hunt to feed their families. The Shoshone attorney testified that his clients passed a game code four years earlier out of necessity to conserve game herds. Both tribes could hunt during the prescribed season under the federal game code, and the Shoshones also had treaty rights they sought to protect by ensuring adequate protection of game. I was not subpoenaed, but Dick testified that hunting regulations were "absolutely necessary."[11]

The controversy was keenly followed statewide and even national media anticipated the precedent-setting legal ruling. Whatever decision Judge Kerr rendered, its impact on resource policy and federal Indian law would be significant. Of course the stakes are always highest close to home. Whatever the outcome, I wondered how it would shape the future of the wildlife and people I had come to care about.

The following week, Judge Kerr ruled in favor of the BIA and Shoshone Tribe. He denied both permanent and temporary injunction requests of the Arapaho Tribe. Kerr reasoned that the Shoshone Tribe would be harmed if the Arapahos continued to hunt virtually unrestricted. His ruling noted that the BIA and Interior Department had authority to enforce the game code under the 1868 Fort Bridger Treaty, which named those entities as trustees of reservation lands. An Arapaho attorney claimed he and his clients were "just shocked by the breadth of ruling." He promised an appeal. It never came.

10

Upshot

I do not think the measure of a civilization is how tall its buildings of concrete are, but rather how well its people have learned to relate to their environment and fellow man.

—Sun Bear of the Chippewa Tribe

At 9,658 feet above sea level, Togwotee Pass was still snowbound on April 29, 2007, as I drove east from Jackson Hole to Lander. Snowmobiles still careened across two to four feet of settled snow now stained with conifer litter and roadside grime that snowplows spewed from the slushy highway. I cracked the car window to inhale the heady scent of fir and spruce. A pine squirrel dashed frantically across the road, a cone clenched in his teeth. I saw no elk tracks punched into the snow. The migration to the high country from winter ranges farther south had not begun.

I had kept in touch with state, federal, and tribal officials for several years after leaving the Wind River. Their reports and newspaper accounts touted visible gains in populations of several species just two to three years after the game code's enactment. The interviews I had scheduled over the next two days would provide more insight a quarter century after I had left.

Dave Skates was the current project leader of the Technical Assistance Office in Lander. He assumed that position after Dick Baldes retired in 1996. Dave was also a fisheries biologist with a relaxed demeanor and an infectious smile. In June 1982, he left his job at the Jackson National Fish Hatchery in western Wyoming for the staff fisheries position in Lander. That same day I left Lander en route to my next career challenge. I likely passed Dave's vehicle headed the opposite direction over Togwotee Pass; two U-haul trailers passing on the Pass.

With Dave and Pat Hnilicka, the current wildlife biologist at the Lander office, I spent the morning catching up on the past 25 years. Pat was a highly competent wildlifer in his thirties who had worked for the Wyoming Game and Fish Department in their Lander office. Pat felt as privileged to work on WRIR as I had. We had communicated occasionally about reservation issues during his six years at the Lander office. Now the three of us had more time to resole the shoe.

Dave recalled that by 1986 or '87, folks began seeing more game animals on the reservation—especially at lower elevations where, prior to the game code's adoption, they had become as rare as manual typewriters at business machine stores. During the turbulent post–game code years, law enforcement efforts focused on relatively few individuals—mostly the same cast of characters who had abused reservation resources previously. By 1988, the BIA enforcement staff was replaced by a cadre of Shoshone and Arapaho wardens, including Gary Lajeunesse and Rawlin Friday. The tribes regained all governing authority of their wildlife management program.

Later that day, I met with Gary and Ray Nation. Ray was now the assistant superintendent of the BIA agency at Fort

Washakie—quite an ascent from biotech at our Lander office. Gary managed the reservation's livestock grazing program, having left the Game and Fish Department nine years earlier. They both declared the game code a success and regaled me with stories of giant mulie bucks and bighorn rams recently taken by tribal members. A 14-year-old, for example, had killed a ram in Wind River Canyon tallying 190 points on the Boone and Crockett scoring system. Production of such trophy animals is one yardstick of game management success to many hunters and biologists.

At the Wyoming Game and Fish Department office in Lander, Regional Supervisor Kent Schmidlin related another telling observation. "Before the game code was enacted," he recalled, "many tribal members bought Wyoming licenses and hunted off the reservation because they had a better chance for success."

The stunning recovery of reservation game herds after 1984 likely cost the state some license revenue.

I spent the evening at Rawlin Friday's home in Ethete. Over bowls of homemade bean soup, he and I reminisced about Blue Sky Hall and counting game from choppers—including our stroll off Trail Ridge in January 1980. We also shared our personal stories from the intervening years and how they had shaped our lives.

"Maybe my toughest decision was in 2006. I left the warden force after twenty-two years of service," Rawley said.

More than his words of regret, the ache in his voice betrayed how he missed the work. Rawley had been a dedicated game warden and took great pride in his profession. He had served longer than any other WRIR warden. But his body wasn't holding up to the physical rigors of the work; he had a bum knee he would later have replaced. In a muted voice he added another reason that made the job even more difficult. In 2005, he investigated the shooting of a bald eagle on the reservation. Because of the bird's federally protected status under the Bald and Golden Eagle Protection Act, he turned over the findings of his investigation to the USFWS. He had known full well that the perpetrator was his nephew, which resulted in family recriminations. Perhaps the ensuing court proceedings prompted his leaving. I didn't care to open old wounds.

After differing rulings by the Tenth District and Appellate federal courts, I learned that the case headed for the U.S. Supreme Court to decide the issue of whether the Religious Freedom Restoration Act allows Native Americans to kill eagles for religious ceremonies without a permit. The Supreme Court declined to hear the case in 2009. That October the case was remanded to the U.S. District Court in Cheyenne, Wyoming, for resolution. There, Judge Alan Johnson agreed to assign the case and prosecution of Friday to the WRIR tribal court.

A parade of kids from elementary to high school age—Rawley's own and their friends—trailed in the front door, to the refrigerator, and then to the adjacent living room. The older ones answered my questions unhesitatingly about their school activities and what they planned to do after graduation. When Rawley walked me to my car, we bear-hugged and vowed to keep in closer contact. As I drove back to Lander, I reflected how long ago it was that I had felt a part of this land and the causes that Rawley and I shared. Renewing this friendship seemed to shrink the intervening decades somehow.

🦌 🦌 🦌 🦌 🦌

The following morning I returned to the Lander Technical Assistance Office. I asked Dave and Pat about current population sizes and harvests of wildlife on the reservation. I was gratified that current numbers approached the projections I forecast in 1982; roughly threefold increases. However, elk populations in the Wind River Range and Owl Creek

Mountains had blown past the optimum numbers. Instead of 3,500, Pat told me that 6,000 to 7,000 elk now roamed the reservation (a number that swelled to 8,000 in 2010).

Table 1. Big game populations on Wind River Indian Reservation in 1982, proposed population size, and numbers estimated in 2007.[1,2]

Species	1982 Population	Proposed Population	2007 Population
Pronghorn	1,000	2,750	2,600–3,100
Bighorn sheep	150	495	350–450
Mule deer	Less than 1,000	3,000	3,200–4,800 (both species)
Whitetail deer	Less than 150	Maintain 150	
Moose	75	240	100–200
Elk	2,550	3,500	5,900–7,100

This stunning achievement was both good news and bad. Just as a successful cattle rancher is first and foremost a good grass farmer, a successful wildlife manager is foremost a good steward of habitat. This requires keeping wildlife numbers in balance with available food and other habitat resources. Consequently, reservation hunting seasons in recent years ran from September 1 to December 31, the longest in the state of Wyoming. Furthermore, since 1989 tribal members could harvest as many as five elk per hunter. Nowhere else in North America that I'm aware of offers such liberal harvests. Formerly ardent opponents of hunting regulations became converts praising their profits.

"Want to hear something interesting?" Dave rhetorically asked. "Jimmy Coulston is now the biggest advocate of the game code!"

All this testified to the success of the reservation's game laws and enforcement efforts. It also signified a supply that exceeded demand. Although the annual elk harvest had jumped from 190 in 1985 to 515 some 20 years later, hunter success remained at 27 percent while elk numbers increased 250 percent. You would expect hunters to welcome such a problem. But I concurred with Dave and Pat that an ever-increasing elk herd was inadvisable. Habitat damage, exacerbated by the ongoing eight-year drought, loomed as a latent by-product.

On the other hand, WRIR's success was a windfall for Wyoming elk hunters. "Substantial numbers of elk migrate from winter range on the reservation and spend summer and fall on adjacent national forest lands," Kent Schmidlin explained as we visited in his office later.

"Some mule deer, bighorns, pronghorns, and moose did the same when I was here," I noted.

"Still do."

"That must provide incentive for the tribes and the state to cooperatively manage those species now that populations are recovered," I added.

"And we do in a number of ways, especially on enforcement issues. Another example is that soon after the game code was put in place, the State provided pronghorns and bighorns to the tribes." Kent spoke enthusiastically about how those animals served as seed stock to reestablish depleted populations.

I felt gratified the wildlife management plan had served as an instrument of wildlife recovery. The seeds of conservation had taken root. Pronghorns and bighorns were flourishing where once they were gone. This restoration continues today. From aerial surveys in 2010, Pat estimated WRIR's migratory pronghorn population exceeded 6,500 animals.

🦌 🦌 🦌 🦌 🦌

At a streamside city park in Lander, I spotted Wes Martel pushing a double-wide blue stroller. "Is that the Bruce Smith I knew way back when?" he chuckled as we heartily shook hands and hugged.

"Same one, just a little older. What do you have here?"

"These are my twin grandsons," Wes beamed.

At one end of a picnic table Wes parked the stroller. Dressed in matching yellow shirts and blue denim jumpers, the fragile cargo locked onto me with four brown eyes. As we visited at just arm's length, I was struck by the relentless passage of time.

When I called Wes to arrange our meeting, it was the first we had talked in 25 years. "So what are you doing these days? Are you still on the Shoshone Council?" I eagerly asked at the sound of his familiar nasal baritone.

"I've been busier than ever, Bruce. I'm no longer on the Council. Gave that up. Now I'm juggling my resource consulting work with taking care of two babies."

Wes was my age. I knew he had grown children. I chose not to ask about the babies now, not knowing where the conversation might lead. That could wait.

"You're probably wondering why I'm calling after all these years, Wes. I'm writing a book. It's about my time at Wind River, the people I worked with, the wildlife, and adoption of the game code," I explained.

"Oh?"

"I'm planning to be in Lander next month. I'd really like to talk with you about your perspectives."

"Sure. That would be good," he replied as an infant fussed in the background.

"Maybe we could meet for breakfast or lunch?"

We set a date and time. I was really looking forward to seeing Wes again.

On my way to breakfast the previous morning, I had picked up a copy of the *Casper Star Tribune*. After a few comments about how the years had treated our boyish looks, I told Wes I was shocked by the paper's lead story, "It's Sad as Hell, Meth Leaves Deep Scars on Reservation." I had incredulously read the tale of how a Mexican drug gang had targeted WRIR in 2000. Initially they offered free methamphetamine to tribal members. The men pursued Indian women, providing them meth as they fathered their children. To support their habit and their children, the women became dealers themselves.[3]

The gang's plan worked insidiously well. Pounds and pounds of meth crossed the Mexican border to Los Angeles, then to Ogden, Utah, where the gang leader lived. The poison ultimately landed in Wyoming. The reservation became awash in meth. Customers became dealers. Crime soared. From 2003 to 2006, cases of child neglect rose 131 percent. Spousal abuse rose 218 percent. Numbers of methamphetamine contacts at the Indian Health Services facility rose 250 percent.

The WRIR was not alone. The BIA recently listed meth as the greatest single threat to Indian communities. Use of methamphetamines is higher among Native Americans than any other ethnic group.[4] Vulnerable due to rural isolation, high unemployment, and understaffed police forces, Indian populations in other states also fell prey to Mexican drug cartels. To stem the tide, several FBI agents were stationed in Lander. The Wind River cartel's leader and 22 others were arrested and convicted in 2005. Despite additional convictions, plenty of dealers and users remained. In 2006, another bust at Wind River resulted in 43 arrests, the largest drug bust in Wyoming history.

A support group for meth-ravaged families, Partners Against Meth, struggles to attract volunteers. As tribal police chief Leon Noseep lamented to the *Casper Star Tribune*, "It's here and it's not going to go anywhere. It's never going to go away."

"My daughter," Wes calmly shared, "is one of the reservation's many victims of meth."

Like the lives of so many others now caring for grandchildren of their meth-addicted children, Wes's life had dramatically changed. Paradoxically, his daughter was fortunate. She had not ended up in jail or worse, as so many others had. "She's living with relatives elsewhere now. We had to get her away from here," he said, referring to the ubiquitous temptations in a land drowning in drugs. "She's doing better. I think she's on the road to recovery."

Overshadowing the newspaper account, his personal story cut to my core. With his grandchildren's liquid brown eyes staring up at me from the stroller, I felt a stab of pain in my chest. Then anger welled at the thought of 30 percent of the reservation's population engulfed in a meth nightmare, destroying lives, homes, and communities. Hardly a reservation family name was spared—all for the drug dealers' profit.

I hadn't known. I had been insulated from such heartbreak, as most easterners had been from the devastation wrought by smallpox on the Plains Indian nations over a century earlier. Although that plague's deliberate spread to American Indians is disputed, there's little doubt of this new epidemic's intentional toll on Indian people.

Stunned by this tragedy, I was surely less comfortable with the conversation than Wes, who lived its reality each day. I offered my heartfelt empathy, but found no other words to say. I awkwardly changed the subject to the reservation's wildlife achievements.

I began by recounting the upsurge in game numbers that Dave and Pat had traced. "This is what you and Frank Enos, Pius Moss, and others told your people that their future could hold."

"I know," he chuckled and shook his head, his long braid of hair swinging freely across his back.

Then his expression turned solemn. A group of children from a reservation school played in the background on swing sets and slides, and scaled a log jungle gym. Their laughter and shrieks pierced the spring air. Wes looked from the kids and two vigilant teachers to a pair of ravens squawking in the newly leafed cottonwoods, and then focused again on me. More clearly than I had heard anyone explain it before, Wes clarified the game code's significance.

"We are connected to everything around us—plants, trees, birds, fish, all the animals. It's all our relation. They all have a spirit like us. We must try to protect our relations as a way of giving thanks for what the Creator has provided us. During our public meetings in the various communities when we were pushing for the game code, some tribal members accused me of acting like a white man—trying to take away their treaty rights. But my response to that was 'what good is a treaty right if there is nothing left to hunt or the state of Wyoming takes control?'"

As witness to the resurgence in wildlife, Wes gushed at what had been accomplished. "Now people who were opposed and called us out during the public hearings are gratified to see the result [of the game code's adoption]. I would say there is 90 percent acceptance."

After pausing in self-reflection, he continued, "Anyway, I am proud to have been part of this effort because I think it is critical for tribes to merge science and technology with culture and tradition. This is the best way to protect our sovereignty and way of life."

I acknowledged the central role he had played in the effort. Then, as a flush of admiring emotion welled in me, I changed the subject again, "So what are you working on nowadays?"

His efforts were twofold. "For one, I'm helping my people hold on to their traditions and language. My twin grandsons were given English names at birth, like other Indian children. Now I'm planning the ceremony where they'll receive their Arapaho names."

As a tribal elder now, Wes mentored others in Arapaho customs. Consecrating Arapaho names on each new child was a vital first step toward ensuring native identity. Second, Wes was facilitating tribal partnering with industry to develop renewable energy resources on the reservation, both to use locally and to market to others. "Just like the game code, we need to connect the science and technical side to the cultural and spiritual side for our people," he explained.

In that moment, I was reminded of the cultural apocalypse that had befallen North America's native peoples. Wes's twin grandbabies were just seven or eight generations removed from Chief Washakie signing the Fort Bridger Treaty in 1863. Ironically, this city park where he (the true native) and I (the descendent interloper) reminisced was part of the 44 million acres of the original WRIR, 42 million of which were purloined from the Indian people over time.

Yet Wes remained a strident optimist, tiptoeing a tightrope between disparate ways of life. As with the reservation's wildlife a quarter century earlier, he focused on possibilities and opportunities for his peoples' future. As he so passionately spoke, his eyes unconsciously wandered to the innocent, round faces beside us.

🦌 🦌 🦌 🦌 🦌

When I began my work at WRIR, the goal as I saw it was to recover the reservation's wildlife. It was about the animals. By the time four years had passed (and increasingly I have realized since), success was about far more—the Shoshone and Arapaho people; the hopeful round faces of kindergartners, the aging elders wanting their heritage secured, and all those in between with a common stake in their reservation's future. In the process some hard hearts were tempered, some skeptics won over, and a measure of reconciliation between two tribes advanced. Unlike my foretaste in California of the necessity of the wildlife-people connection—there to facilitate the execution of conservation—I now understood people were the *object* of the work as much as restoring wildlife for wildlife's sake. My greater mission at WRIR had been to help fortify the Shoshones' and Arapahos' cultural link to the land and to their past.

As I listened to Wes Martel, a deep reaction welled inside me. I had momentarily drifted to distant conversations with Herman Lajeunesse, Pius Moss, Jess Miller, Art Nipper, and others long since passed from this place. I yearned to hear their stories once again, see their eyes dance as they talked about the land and the wild things held close in their hearts. But I was thankful, too. Before each had walked on the wind to the spirit world, he had seen the Creator's gifts begin to return to the people of Wind River. As in most indigenous cultures, wildlife was a link from the past to the present and into the future. The Plains Indian not only ate buffalo, deer, and elk; the animals provided his housing, dress, arts, and religion, his history and conclusion—a living bridge to his ancestors and the spirit world beyond.

The game code's advent proved a remarkable achievement. It benefited wildlife and the people who treasured it, both on and adjacent to the reservation. That triumph of conservation, as contentious and protracted as the struggle was, serves as a model, a ray of hope, for other reservations and other native peoples seeking to preserve their heritage. In a sense, the WRIR was a microcosm of the worldwide challenge of rescuing and sustaining the earth's biodiversity—something renewable yet fragile enough to slip into oblivion like the passenger pigeon or Tasmanian wolf. If wildlife could be recovered here, it could be restored elsewhere. The battle was symbolic as well, I realized while listening to Wes, of two nations seeking to hold on to their past yet provide for their future.

Someone once asked me what the secret was to the WRIR's conservation success. "There's no secret to success." I replied. "Did you ever know a successful person who didn't tell you all about it?" This was an achievement in which I was one of many proud participants. This was a story whose success begged to be told.

Epilogue

A land ethic ... reflects the existence of an ecological conscience, and this in turn reflects a conviction of individual responsibility for the health of the land. Health is the capacity of the land for self-renewal. Conservation is our effort to understand and preserve this capacity.

—Aldo Leopold, *A Sand County Almanac* (1949)

Traditional beliefs of native people regard all life, including humans, as connected on a parallel plane. This horizontal organization incorporates spirituality without hierarchy, arrogance, or subservience. There is "other" power, but not "higher" power, with the spirit of Creation infused in all things. Such is the source of native peoples' reverence for all Nature.

Some believe today's great moral issue is to take care of each other. Or as the fourteenth Dalai Lama succinctly asserted, "My true religion is kindness." I believe kindness extends to caring for all that enriches our lives in the world around us, including wild things and wild places. I can think of no more important undertaking than ensuring a future for all life on this planet. Regardless of where we live, our ethnicity, religion, social or economic status, humans cannot persist in a world that cannot sustain life. Because of humankind's profound influence on the environment, we can no longer idly expect Nature to independently heal its wounds. So what is our role?

Professional resource managers are increasingly overwhelmed by the challenges of environmental protection. Their ranks are not keeping pace with human population growth and demand on the land's resources. Based on a 2004 survey of more than 5,000 employees of 39 state agencies working in the fields of fisheries and wildlife management, law enforcement, and information and education, 47 percent said they planned to retire by 2015. Even more dramatic were the expected retirements of 77 percent of those in leadership positions.[1] This pattern is similar in federal government positions. Why? Primarily the exodus of baby boomers like myself. Many in my profession fear that the erosion of an experienced workforce in the field of wildlife management and conservation may be irreversible. At fault are shrinking budgets for natural resource programs, decreasing enrollment of students in traditional wildlife and fisheries management programs, and a pendulum shift in academia toward nonapplied fields such as conservation biology and environmental science. Utah State University wildlife professor Steve McMullin notes that even those students who take traditional wildlife biology courses aren't necessarily going to consider working for federal or state wildlife agencies when they graduate. Increasingly, students choose to work for wildlife rehabilitation centers, zoos, and nonprofit organizations. "It's maybe one or two out of ten students who express interest in management agencies as a career," McMullin said.[1]

With many universities now adopting a more theoretical rather than an applied focus, there is a dwindling supply of those trained in the practice of studying, managing, and conserving wildlife on the ground. Beyond field biologists and resource managers, the ranks of conservation educators are also thinning. With so many people now living in urban or suburban environments and unfamiliar with the natural world, teaching an appreciation of Nature and wildlife is an increasingly worthy and weighty task.

In a BBC interview in 2005, the eminent natural history film maker Sir David Attenborough condensed the importance of nature education this way: "We'll only get people to care for the environment if they know something about it."[2] This too has been the battle cry of the Harvard University scientist E. O. Wilson. "Education in biology is important not just for the welfare of humanity, but for the survival of the rest of life."[3]

For those who would choose the challenge of a career as a wildlifer or other natural resource professional, your choice is a noble one. However, conservation of public resources is a shared responsibility. All citizens own and benefit from healthy landscapes and a vibrant diversity of life—not just those residing in rural communities or near wildlands like the WRIR.

Reserving a place for wildlife requires public support and stewardship. That takes people—lots of us. Legions of citizen conservationists in concert with scientists must advocate and attend to wildlife's future. Citizen conservation is not only stimulating and spiritually satisfying, it's a precious reward of free societies where wildlife and other resources are owned in common. Each of us can make a mark in a multitude of ways: join a conservation organization or wildlife club, make informed political choices, support environmental causes, and practice lifestyles compatible with communities of wild things. "Individual commitment to a group effort—that is what makes a team work, a company work, a society work, a civilization work," noted NFL football coach Vince Lombardi.

Recovery of WRIR's wildlife faced significant challenges: intertribal conflicts, defense of dearly held treaty rights, hard-earned suspicion of the white man's ways, and all humanity's intrinsic resistance to change. It is to their credit that the Shoshone and Arapaho people overcame these obstacles. But without a few prominent tribal leaders, elders, and other advocates, an increasingly bleak future faced wildlife on the Wind. Besides those leaders undaunted by criticism and unfailing in their dedication, this was an accomplishment of many. The names of those who spoke to friends and family about the need for conservation would easily fill this page. Their collective efforts and perseverance will benefit generations to come.

With populations of all other large mammals, even formerly extirpated grizzly bears and gray wolves, prospering or on the road to recovery, only one species remains notably missing on WRIR—the bison. That too may one day change. Officials from Yellowstone National Park and the state of Montana began trapping bison on the park's northern and western borders several years ago. To limit numbers and the chance that bison might infect cattle in Montana and Idaho with the disease bovine brucellosis, captured bison that tested positive for the disease were destroyed. Rather than destroy all surplus animals that were captured, a quarantine facility was constructed north of Yellowstone in 2004. Young bison trapped since 2005 have been held and retested. Those that have repeatedly tested negative for brucellosis may provide a source for new bison herds with the Yellowstone population's unique gene pool.

Leadership of the Northern Arapaho Tribe requested 40 bison from the state of Montana in early 2009. The tribal council's letter explains the appeal: "traditional lives will again have a balance, a traditional source of buffalo meat and artifacts as opposed to the search and buy method used today for needed ceremonial items."[4] A planned 32,000-acre pasture west of Wind River Canyon would provide a halfway house for a free-range herd. However, fearing the Yellowstone bison might infect their cattle with disease, the Arapaho General Council vetoed the endeavor. Still, a tribal homeland rich with bison may lie somewhere down the road. Should this transpire it won't restore the nineteenth century buffalo economy of two Plains Indian nations. But restoring Nature's wholesome condition brings honor to those who undertake the effort. Each step to return wildlife to the Wind helps complete the community of life of the Arapaho and Shoshone people.

Acknowledgments

Writing this book is something I have always known I would do. If I had stayed longer, more than the blink-of-an-eye four years I was at Wind River Indian Reservation, I may have completed this much sooner. But that was not to be. After my transfer from the U.S. Fish and Wildlife Service's Technical Assistance Office in Lander, Wyoming, in June 1982, my new assignment at the National Elk Refuge in Jackson Hole, Wyoming, distracted my attention and occupied my time. The book incubated for another 22 years until I retired from federal government service—a long enough interlude that I could have lost interest. But I never did. The wildlife, the people, and the story fermented in me. Fortunately, when I set to the task of resurrecting the past, my pack rat nature helped refresh that history. Reports, correspondence, press clippings, taped interviews, and notebooks provided the bones of the narrative. But those alone were not enough. I corroborated my memories of events with others who had shared them. And for the book's final chapter, "Upshot," I wanted to recount how events after my departure from Wind River Reservation had played out. For both purposes I turned to reliable colleagues and friends who generously shared their time and recollections with me. In most cases, my memories concurred with theirs. When slight conflicts occurred, I chose whoever's recollections seemed clearer. I hope I have done justice to those events from three decades ago and to the people whose lives they touched. Any factual errors in recounting details of past events are mine alone.

I had the privilege of working with many Shoshone and Arapaho councilmen and game wardens who were dedicated to the restoration of their wildlife heritage. Foremost among them I thank Rawlin Friday, Gary Lajeunesse, Wesley Martel, Bob St. Clair, and the late Dr. Frank Enos, Robert Harris, Sr., Pius Moss, Emil O'Neal, and Ernest SunRhodes. The vision and dedication of each of these individuals gave reservation wildlife a new chance.

For their willingness to endure hours of grilling about reservation history I thank each of the following: Jim Barquin, Bill Bradford, Sr., Pearly "Junior" Brooks, Bill Brown, Jim Fike, Ben Friday, Sr., Raymond Harris, Jim Hill, Burke Johnson, Herman Lajeunesse, Leo Lajeunesse, Bud LeClair, Sr., Dutch McAdams, Jesse Miller, Pius Moss, Art Nipper, Jr., Carl Shatto, Martha Stagner, Levi Surrell, Ernest SunRhodes, and Landis Webber. Many of these tribal elders and reservation "old-timers," whose memories I tapped to reconstruct wildlife's history, are no longer with us (now part of the Great Mystery). Only a sketch of the knowledge they shared appears in the pages I have written here. That I captured these men and women's oral history is testament to their desire to leave a record of what touched their lives and was instructive to pass on to others. Edith Johnson and Shirley Enos provided photographs of deceased loved ones whose visages live on in this book.

For carrying out fieldwork, including wildlife surveys and habitat inventories, and assisting with report preparation, I give my heartfelt thanks to Dick Baldes, Kevin Berner, Paul Bovitz, Todd Day, Rawlin Friday, Judy Gillette, Connie Isdahl, Edith Johnson, Richard Johnson, Gary Lajeunesse, Philip Mesteth, Ray Nation, Betty Parrish, Reg Reisenbichler, and Dave Vogel. These individuals helped produce an amazing baseline of information about wildlife populations and habitats within Wind River's 2.2 million acres. Their tireless dedication to conservation's cause I won't forget.

During hundreds of hours of wildlife surveys I was safely piloted by Ron Gipe, Larry Hastings, Bob Hawkins, and Dan Hawkins. I benefited from the knowledge of other federal and state biologists and game wardens who shared their knowledge with me: Bill Crump, John Emmerick, Bill Eversole, Pat Hnilicka, Marvin Hockley, Rich Olson, Bob Oakleaf, Ken Persson, Kent Schmidlin, Leonard Serduik, and Dave Skates.

Publishing a book is no simple matter. Those who reviewed some or all draft chapters include: Bill Alldredge, Don Burgess, Carol Cunningham, Ted Kerasote, Carl Mitchell, and Meredith and Tory Taylor. Each provided insights and comments that strengthened my writing. I owe particular thanks to Don and Ted. As an exceptional editor, Don advised reorganization of portions of the text. Furthermore, he suggested I reconsider my original intent of covering my complete government career in one book. After reading nearly 300 pages of text— just half of the anticipated length—I can't imagine how he came to that conclusion! It was obvious to him that my 22 years in Jackson Hole needed to wait for a future airing. He was right. The Wind River's story deserved to stand on its own.

Ted Kerasote was a reliable mentor, walking me through the labyrinths of the book publishing business based on his own trial and error experiences. His sage advice helped get this to press.

My first book, *Imperfect Pasture: A Century of Change at the National Elk Refuge in Jackson Hole, Wyoming*, documenting habitat changes and cascading ecological effects on wildlife, was published by a specialty publisher. To reach a broader audience with this book, I sought an academic press. I was fortunate that editor John Alley of Utah State University Press believed in the project. His guidance and the skill of the press's production manager, Dan Miller, brought this story to life, producing a product of which we're all proud.

From the first keystrokes at the computer to the book's appearance in print, my wife Diana was my constant sounding board and my work's biggest fan. When I wondered if what spilled onto paper would interest others, she always agreed to read another page. Love will do that, I guess. I love you my dear.

Finally, I don't pretend to speak for American Indians. My words about the Shoshone and Arapaho people are my interpretations and impressions from literature research and my experiences among them. As for direct quotations attributed to others, I have largely limited those to tape-recorded interviews, personal notes, reports, and newspaper articles. Yet there are bits of conversation that are not word for word. In those instances, I have recreated exchanges as best I can recall and accurately captured the emotional dialogue. For any perceived misrepresentation, I take full responsibility.

References

Introduction

1. American Museum of Natural History. 1998. National survey reveals biodiversity crisis. http://www.well.com/user/davidu/amnh.html

Chapter 1: Gettin' There

1. Dolin, E. J. 2003. *National wildlife refuges.* Washington, DC: Smithsonian Institution Press.
2. Meacham, Jon. 2008. *American lion: Andrew Jackson in the White House.* New York: Random House.
3. U.S. Fish and Wildlife Service. 2000. *Recovery plan for bighorn sheep in the peninsular ranges, California.* Portland, OR: US Fish and Wildlife Service.
4. Hewett, R. J., W. W. Logthouse, and E. L. Sutton. 1977. *Review of Fish and Game Department, Joint Business Council Shoshone-Arapaho Tribes, Wind River Indian Reservation, Fort Washakie, Wyoming to determine problem areas, evaluate certain procedures, and to make specific recommendations.* Washington, DC: Agricultural Resources Corporation of America.
5. Kellert, S. R. 1997. *Kinship and mastery: Biophilia in human evolution and development.* Washington, DC: Island Press.
6. Orwell, G. 1982. *Animal farm.* New York: Harcourt Trade Publishers.
7. Morrison, J. C., W. Sechrest, E. Dinerstein, D. S. Wilcove, and J. F. Lamoreux. 2007. Persistence of large mammal faunas as indicators of global human impacts. *Journal of Mammalogy* 88:1363–1380.

Chapter 2: On the Reservation

1. Ehrlich, G. 1985. *The solace of open spaces.* New York: Penguin.
2. Stamm, H. 1999. *People of the Wind River.* Norman: University of Oklahoma Press.

3. Murray, L. 1978. *The Wind River Reservation yesterday and today.* Riverton, WY: Western Printers.
4. Murphy, R. F., and Y. Murphy. 1960. Shoshone-Bannock subsistence and society. *Anthropological Records* 16(7):360–365.
5. Bonnicksen, T. M. 2000. *America's ancient forests: From the Ice Age to the age of discovery.* New York: John Wiley and Sons.
6. Trenholm, V.C., and M. Carley. 1969. *The Shoshonis: Sentinels of the Rockies.* Norman: University of Oklahoma Press.
7. Billard, J. B., ed. 1974. *The world of the American Indian.* Washington, DC: National Geographic Society.
8. Friederici, P. 2006. *Nature's restoration.* Washington, DC: Island Press.
9. Demarais, S., K. V. Miller, and H. A. Jacobson. 2000. White-tailed deer. In *Ecology and management of large mammals in North America,* ed. S. Demarais and P. R. Krausman, 601–628. Upper Saddle River, NJ: Prentice Hall.
10. Smith, B. L. 1982. *A plan for management of wildlife on the Wind River Indian Reservation.* Lander, WY: US Fish and Wildlife Service.
11. Errington, P. L. 1957. *Of men and marshes.* Ames: Iowa State University Press.
12. Darwin, C., and F. Darwin. 2008. *The autobiography of Charles Darwin.* Amazon.com: CreateSpace.
13. Wilson, E. O. 1995. *Naturalist.* New York: Warner Books.

Chapter 3: First Elk

1. Dearing, M.D. 1997. The function of hay piles of pikas. *Journal of Mammalogy* 78(4): 1156–1163.
2. U.S. Fish and Wildlife Service. 2009. 90-day finding on a petition to list the American pika as threatened or endangered with critical habitat. *Federal Register* 74(87): 21301–21310. May 7, 2009.
3. http://www.fws.gov/mountain-prairie/species/mammals/americanpika/02052010FRTemp.pdf
4. Smith, B. L. 1980. *A preliminary report on the status of elk on the Wind River Indian Reservation.* Lander, WY: US Fish and Wildlife Service.
5. Darwin, C. 1859. *On the origin of species by means of natural selection.* London: Murray Press.
6. Knight, R. L. 1970. The Sun River elk. *Wildlife Monograph* Number 23.
7. Smith, B. L., and K. L. Berner. 1982. *The history, current status, and management of deer on Wind River Indian Reservation.* Lander, WY: US Fish and Wildlife Service.
8. Smith, B. L. 1982. *The history, current status, and management of bighorn sheep on Wind River Indian Reservation.* Lander, WY: US Fish and Wildlife Service.

9. Thorne, T., G. Butler, T. Varcalli, K. Becker, and S. Hayden-Wing. 1979. *The status, mortality and response to management of the bighorn sheep of Whiskey Mountain*. Wildlife Technical Report 7. Cheyenne: Wyoming Game and Fish Department.

Chapter 4: Mountains and Sky

1. Schlegel, M. 1976. Factors affecting calf elk survival in Northcentral Idaho—a progress report. *Proceedings Western Association of State Game and Fish Commissioners*. 56:342–355.

Chapter 6: The Way It Was

1. Murphy, R. F., and Y. Murphy. 1960. Shoshone-Bannock subsistence and society. *Anthropological Records* 16(7):360–365.
2. Stamm, H. 1999. *People of the Wind River*. Norman: University of Oklahoma Press.
3. Murray, L. 1978. *The Wind River Reservation yesterday and today*. Riverton, WY: Western Printers.
4. Smith, B. L. 1981. *The history, current status, and management of pronghorn antelope on Wind River Indian Reservation*. Lander, WY: US Fish and Wildlife Service.
5. Smith, B. L., and K. L. Berner. 1982. *The history, current status, and management of deer on Wind River Indian Reservation*. Lander, WY: US Fish and Wildlife Service.
6. King, C. L. 1963. *Reestablishing the elk in the Bighorn Mountains of Wyoming*. Cheyenne: Wyoming Game and Fish Department.
7. Smith, B. L. 1980. *A preliminary report on the status of elk on the Wind River Indian Reservation*. Lander, WY: US Fish and Wildlife Service.
8. Smith, B. L. 1982. *The history, current status, and management of bighorn sheep on Wind River Indian Reservation*. Lander, WY: US Fish and Wildlife Service.
9. Buechner, H. K. 1960. The bighorn sheep in the United States, its past, present, and future. *Wildlife Monographs* Number 4.
10. Honess, R. F., and N. M. Frost. 1942. A Wyoming bighorn sheep study. *Wyoming Game and Fish Department Bulletin*. Cheyenne: Wyoming Game and Fish Department.
11. Smith, B. L. 1985. Moose and their management on Wind River Indian Reservation. *Alces* 21: 359-391.
12. Diamond, J. 2005. *Collapse: How societies choose to fail or succeed*. New York: Viking

13. Smith, B. L. 1982. *A plan for management of wildlife on the Wind River Indian Reservation.* Lander, WY: US Fish and Wildlife Service.
14. U.S. Fish and Wildlife Service. 2002. *2001 National survey of hunting, fishing, and wildlife-associated recreation.* Washington, DC: US Department of Interior.
15. National Shooting Sports Foundation. 2002. *Families afield.* Newton, CT.

Chapter 7: Younger Kids

1. Wilson, E. O. 2006. *The creation: An appeal to save life on Earth.* New York: W. W. Norton and Company.
2. Louv, R. 2005. *Last child in the woods: Saving our children from nature-deficit disorder.* Chapel Hill, NC: Algonquin Books.
3. Maffly, B. 2008. Playing it too safe? *Montana Outdoors* 39(2):22–27.

Chapter 8: On the Same Page

1. Smith, B. L., and K. L. Berner. 1982. *The status and management of waterfowl on Wind River Indian Reservation.* Lander, WY: US Fish and Wildlife Service.
2. Smith, B. L. 1981. *The history, current status, and management of pronghorn antelope on Wind River Indian Reservation.* Lander, WY: US Fish and Wildlife Service.

Chapter 9: Game Code

1. Tribal viewpoint. 1980. *Wind River Journal,* October 3.
2. Neal, D. 1981. Decline prompts appeal for game management on the reservation. *Riverton Ranger,* May 15.
3. Jones, C. 1982. Wild game on the table today, none on the range tomorrow? *High Country News* 14(3):1, 10–11.
4. United Press International. 1983. Big game slaughter upsets tribal leaders. *Casper Star Tribune,* December 9.
5. Adams, T. 1983. Tribes' elders call for control. *Riverton Ranger,* December 9.
6. Associated Press. 1983. Reservation game code advocated. *Casper Star Tribune,* October 10.
7. Leopold, A. 1949. *A Sand County almanac.* New York: Oxford University Press.

8. Melnykovych, A. 1983. Shoshones seek game moratorium. *Casper Star Tribune*, December 15.
9. Williamson, L. 1984. No laws for the Indians. *Outdoor Life*, May, 33–35, 144.
10. Kier, B. 1984. Indian plans to hunt despite code. *Casper Star Tribune*, October 12.
11. Herman, M. 1984. Hunting curbs necessary to save wildlife from extinction. *Denver Post*, October 26.

Chapter 10: Upshot

1. Smith, B. L. 1982. *A plan for management of wildlife on the Wind River Indian Reservation*. Lander, WY: US Fish and Wildlife Service.
2. U.S. Fish and Wildlife Service and the Eastern Shoshone and Arapaho Tribes. 2007. *Wolf management plan for the Wind River Indian Reservation*. Lander, WY: US Fish and Wildlife Service.
3. Wagner, A. 2007. "It's sad as hell." Meth leaves deep scars on reservation. *Casper Star Tribune*, April 30.
4. National Congress of American Indians. 2006. Methamphetamines in Indian country: An American problem uniquely affecting Indian country. http://www.ncai.org/ncai/Meth/Meth_in_Indian_Country_Fact_Sheet.pdf

Epilogue

1. Unger, K. 2007. The graying of the green generation. *The Wildlife Professional* 1(1):18–23.
2. British Broadcasting Interview of Sir David Attenborough. 2005. State of the Planet. http://www.bbc.co.uk/nature/programmes/tv/state_planet/attenborough.shtml
3. Wilson, E. O. 2006. *The creation: An appeal to save life on Earth*. New York: W. W. Norton and Company.
4. Trosper, K. Letter dated February 9, 2009, from the Northern Arapaho Tribe to the State of Montana.

Index

A

Absaroka Range, 35
Africa, 121, 139
Agricultural Resources Corporation of America, 18
airplanes: Cessna, 49–51, 66–68, 77, 102, 118, 174, 182; Citabria, 50; flying in, 50, 51, 52, 67; Piper Cub, 50; Super Cub, 50
Algonquian, 88
American Indian Movement, 190
American Museum of Natural History, xi
Anasazi, 37, 133
Appleby, Charlie, 131
Aramajo, Chester, Jr., 190
Arapaho Business Council, 184
Arapahoe, WY, 23, 134, 136
Arapahoe Elementary School, 150
Arapaho Entertainment Committee, 183
Arapaho General Council, 178, 181, 183, 184, 186, 189, 190, 210
Arapaho Indians, xi, xii, 13, 19, 23, 30, 31, 55, 74, 96, 115, 122–125, 128, 131, 136, 141–144, 159, 176, 178, 180, 183–185, 187–191, 210, 211, 216; culture, 30, 122, 127, 133, 144, 164, 203, 204; history, 14, 24, 27–29, 36, 124, 148
assimilation, 148
Attenborough, Sir David, 209
attention deficit disorder, 148
Ault, Megan, 148
Aztecs, 121

B

baboons, chacma, 140
backpacking, 63, 75, 77
badger, 44, 163
Baggs, Frank, 134
Bald and Golden Eagle Protection Act, 196
Baldes, Dick: and councilmen, 15–17, 31–33, 160, 164; and educational outreach, 145–146, 179–180; and elders, 122–123; and elk overkill, 186; and FBI, 171; and fisheries, 13, 52; and flying, 49–50; and game code, 177–178, 187, 191; 66; and General Council, 181–184 and horse–packing, 52–55;
Barquin, Jim, 125, 135, 214
bear, black, 64, 152; behavior, 81, 116; description of, 79–80, 82, 126; predation, 83, 86
bear, grizzly, 64, 127, 138, 164, 210; behavior, 81–83, 116; persecution of, x, 83
bear, polar, 59
bears, xii, 20, 22; and Indians, 83; short–faced, 139
Bedford, Clay P., 143
behavior, x, 9, 21, 41, 44, 87–89, 137, 143, 163
Bergstrom, Dr. Robert, 73
Berner, Kevin, 38, 162, 170, 183, 214
bighorn sheep, ix, xii, 32, 66, 129, 138; demise of, x, 131–133; desert, 5–6, 8–12; diseases of, 73, 132–133; evolution, 35; food habits of, 71; hunting of, 37, 124, 131–132,

Index 221

135, 176, 195; and livestock, 71, 131–133; migration, 129–131, 198; occurrence on WRIR, 17, 37, 130–132; population size and management, 70, 185, 187, 197–198t; surveys of, 96–98, 106; translocation of, 70, 198; use by Indians, 76, 124, 133; winter range, 50, 71, 98, 130, 132
Big Robber, 62
Big Sunday Creek, 65
Billings, MT, 144, 160, 162, 185
birds: falcons, xii, 78, 162, 164, 167; and Indians, 202, 164; nongame, 162; of prey, 44, 51, 164, 166–167; rehabilitation of, 168–170; species of WRIR, 17, 46, 141, 162–163; upland game, 135, 176, 190; waterfowl, 78, 162, 176, 190
biscuitroot, 82
bison (buffalo): brucellosis, 210; destruction of, x, 19–20, 26–28, 36, 128, 138, 180; economy, 24, 115, 137, 211; former numbers, 138; hunting of, 24, 27, 115, 124; Indian beliefs about, 128, 133, 137, 204; migrations, 28, 115; restoration, 115, 210–211; use by Indians, 24–26, 36, 76, 120, 124, 133, 138, 210
bitterbrush, 41, 133
Bitterroot Mountains, 79
blackbird, redwing, 42–44
Black Coal, Chief, 29
black-footed ferret, 72, 164, 165
Black Mountain, 96, 115, 130
Black Ridge, 96, 98, 99
bluebird, mountain, 44
Blue Sky Hall, 179–181, 185, 187, 195
Botswana, 139
Boysen Dam, 34
Boysen Reservoir, 34
Bradford, Bill, Sr., 125, 214
Britt, Tim, 172–174
Brooks, Pearly "Junior", 125, 135, 214
budget, 49, 66, 69, 160, 161, 162, 208

Buffalo Eaters. See Kutsindŭka, 26
Bull Lake Creek canyon, 77–80, 83–86, 127–128, 152, 175
Bureau of Indian Affairs (BIA), 32, 39, 46, 163, 189, 191, 194, 200
Bureau of Land Management (BLM), 7–12, 19, 174
Bureau of Reclamation, 163

C

California, xi, 3–4, 7–12, 19, 32, 59, 67, 76, 120, 204
Casper Star Tribune, 188, 199, 200, 220, 221
cattle, 14, 16, 77: brucellosis in, 210; diets of, 71; habitat use by, 41, 163; numbers of, 41; ranchers, 125, 132, 197
ceded reservation lands, 23, 124, 125, 135
cheetah, 35
Cheney, Congressman Dick, 188
Cherokee Trail of Tears, 121
churches and missions, 4, 120
Churchill, Winston, 183
Clark, Secretary of the Interior William, 188
climate change, 46, 59
Clovis, 139
Colorado State University, 71
Columbia River Plateau, 24
Comanche Indians, 24
Comprehensive Employment and Training Act (CETA), 175
conservation, ix–xi, 7, 11–13, 21, 123, 140, 161, 164, 177; citizen, 147–148; education, 150; ethic, 144, 148; and game code, 145, 184, 189; laws, 189; plan, 18, 30, 149; principles, 18, 30, 149; success, xii, 198, 204–205
Cortés, Hernán, 121
Coulston, Jimmy, 170, 171, 175, 183, 184, 197
coyote, 37, 45, 51, 136, 173
Creator, xii, 188, 202, 204. See also Great Spirit

Cretaceous Period, xi
Crow Creek, 96, 98, 100, 101, 108, 109, 112–114, 130, 138
Crowheart, WY, 23, 164
Crowheart Butte, 62, 102
Crow Mountain, 117, 130
culture: cleansing of, 121, 148, 203; Clovis, 139; and conservation, 18; distinctions, 14, 15, 122; heritage, x, 138, 140; and Nature, 140, 144, 147, 204; Shoshone, 37, 76; and science, 202–203; Tubatulabal, 76; understanding of, 122, 140, 150; and wildlife, xii, 133, 144, 21, 164, 204

D

Dalai Lama, 207
Dawes, Senator Henry, 28
Day, Tad, 38, 45, 46, 63, 65, 82
deer: distribution and habitat of, 36, 41, 50; food requirements, 36, 38, 71, 152; and game code, 135, 176, 178, 190; hunting of, 18, 76, 124, 133, 177, 186, 188; importance to Indians, 36, 133; Indian beliefs about, 133, 204; management of, 185, 187; occurrence on WRIR, 17, 115, 138; populations on WRIR, 38, 125, 135–136, 187; study of, 4–5, 20, 46, 51; surveys of, 67, 70, 92–93, 96, 98; swimming, 78. *See also* mule deer, white–tailed deer
Desert Land Entry Act, 10
Diamond, Jared, 140
Dinwoody Ridge, ix, 71, 75, 127
Dioum, Baba, 149
disease: of Indians, 28, 133, 201; of wildlife, 28, 73, 133, 210
Dry Creek canyon, 86
ducks, ix, 78
Dukarika. *See* Sheep Eaters
DuNoir Wilderness, 128

E

eagles, 5, 51, 72, 120, 140, 153, 173, 196: bald, 164, 195; golden, 73, 96, 164

Eastern Shoshones, *See* Shoshone Indians
East Fork of the Wind River, 107, 108, 113
education, 21, 121, 143–154, 208, 209
elk: adaptations, 116; aerial surveys of, 38, 47–50, 66–70, 86, 91–95, 97, 160; calves, 60, 64, 65, 77–86; classifications of, 67–69; distribution, winter 36–38, 41, 45, 49, 70, 129, 136, 160, 198; evolution of, 35; feces, 41, 60, 70, 78; food supply, 38, 40; forage requirements, 36, 38–39, 40, 71, 114; habitat use of, 41, 76, 123; habitat mapping, 63–65, 80; historical abundance, 27, 125, 127–128, 135, 136, 139; hunting of, 18, 135, 176–178, 186–188, 190, 197–198; importance to Indians, 17, 34, 124, 133, 204; and logging, 63–66; on Lolo National Forest, 62–66; migration, 28, 68, 128, 129, 198; in Montana, 20, 62, 70; mortality, 69; nursing, 60, 84; population estimates, 38, 40–41, 69–70, 187, 196–197t; in roadless areas, 62–66; sounds and smells of, 65, 84; transplanting, 128; Tule, 75; and weather, 68, 69, 186
emergency locator transmitter (ELT), 105–107
Emmerick, John, 87, 215
Endangered Species Act, 164
energy production, 11, 73, 203
Enos, Frank, 16, 19, 122–123, 144, 161, 176–177, 201
enrollment, 14, 18, 30, 31, 136, 137, 141, 177, 181, 182, 187, 208
Errington, Paul, 43
Eurasia, 35, 139
Euro–Americans, x, 3, 10, 121, 124, 132, 138–140, 148
evolution, x, 21, 35, 41, 139, 188
extinction, xi, 35, 59, 83

F

farming, xi, 28, 29, 36, 135
Federal Land Policy and Management Act, 11
Felter, Wayne, 186
fences, 115, 172–175
Fike, Jim, 130, 214
Fish and Game Committee, Tribal, 19, 67, 141, 145, 159, 174, 176, 180, 190
Fish and Game Department, Tribal, 18, 19
fish: fisheries, 13, 52, 194, 208; fishing, 30, 43, 55
flying. *See* airplanes and helicopters
food habits of wildlife, 38, 40, 41, 58, 59, 71, 72, 77, 82, 88, 132, 197. *See also* specific species
forage supply, 38, 40–42
Fort Bridger Treaty, 26, 27, 191, 203
Fort Washakie, WY, 15, 24, 31, 125, 167, 171, 194
Friday, Rawley (Rawlin), 96–101, 103–105, 107–114, 116–118, 128, 194, 195
Friederici, Peter, 29

G

Game and Fish Department, Wyoming, 36, 87, 134, 165, 194, 195
game code: development of new, 136, 145–146, 187–191, 193–195, 197–199, 202–204; 1948–1953 code, 135–136, 176
Gas Hills Highway, 172
Gavin, John, 177
General Allotment Act, 28
General Council, Arapaho, 30, 149, 178, 181, 183, 184, 186, 189, 190, 210
General Council, Shoshone, 30, 176, 177
General Land Office, 10
Geologic Time: Cretaceous Period, xi; Jurassic Period, 76; Little Ice Age, 24; Pleistocene, 35; Pre-Cambrian, 129; 139, 140; Triassic Period, 130
Gipe, Ron, 92, 103, 215
global warming. *See* climate change
Grand Teton National Park, 23
grasshoppers, 167–169
Great Basin, 24, 59, 76
Greater Yellowstone Ecosystem, 127
Great Plains, 24, 26, 72, 115
Great Plains Hall, 179
Great Spirit (Great Mystery), 1, 120, 214
Gros Ventre River, 134
guns, 24, 26, 92

H

habitat: carrying capacity, 36, 143; management plan, 8, 19; riparian, 41, 88, 134, 162; sampling, 38, 39, 42, 44, 45, 170; wetland, 43, 162–163
Halvorson, Gary, 62, 63, 65, 66
Harris, Bob (Robert), Sr., 15–18, 32–33, 132, 160, 186, 188, 190
Harris, Fred, 136
Hastings, Larry, 66, 67, 102, 174, 215
Hawkins and Powers Aviation (H & P), 107, 119
Hawkins, Dan, 119, 215
Heebeecheeche, Lake, 75
helicopters: Bell 47, 92, 95; Bell and Hiller, 67, 68; Bell Jet Ranger, 118; CH–46 Sea Knight, 94; Cobra, 93; Hiller 12E, 96; Huey, 93
heritage, cultural, x, 140, 188, 203, 204
High Country News, 184
Hill, Jim, 109, 128, 214
Hiller, Stanley, 97
Hnilicka, Pat 194, 196–198, 201, 215
homeland, 3, 24, 27, 66, 86, 211
Homestead Act, 10
Hornaday, William, 28
horses: Indian acquisition of, 24, 26; packing, 52–56, 75; feral, 41, 71
hunter–gatherer lifestyle, 26, 28, 140, 148

hunting: by Indians, 24, 30, 37, 62, 124–125, 133, 135, 142–144, 176–177; licenses, 175, 178; market, 132; moratorium on, 187–189; regulation of, xii, 15, 18–19, 29, 135–137, 141, 145, 163, 170, 178, 181, 186–187, 190–191, 197; state regulations, 124–125, 128, 135; rights, 30, 183, 188

I

Indian Health Services, 200
Indian Removal Act, 10
Indian tribes: Apache, 4; Blackfeet, 26; Cherokee, 4, 121; Cheyenne, 24; Comanche, 24; Crow, 62; Dakota, 24; Iroquois, 3; Kiowa, 120; Nez Perce, 121; Sioux, 1, 3, 46

J

Jackson, President Andrew, 10
Jackson Hole, WY, 23, 89, 115, 127, 128, 193, 213, 215
Jackson National Fish Hatchery, 194
jaguar, 22
Johnson, Burke, 130, 131, 214
Johnson, Judge Alan, 196
Johnson, Samuel, 180
Joint Business Council, 18, 30, 31, 32, 67, 70, 91, 141, 159, 161, 164, 174, 186, 190
Joseph, Chief, 121

K

Kate's Basin, 94
Kellert, Stephen, 21
Kern River Plateau, 76, 77
Kerr, Judge Ewing, 191
kestrel, American, 166–170
kindergartners, 151, 154, 203
King, Cal, 128
Knight, Dick, 70
Kutsindüka, 26

L

Lajeunesse, Gary, 127, 186, 194, 195, 214

Lajeunesse, Herman (wife Rachael), 125, 126, 135, 204, 214
Lander, WY, 6, 13, 15, 19, 23, 45, 68, 87, 92, 105, 131, 149, 150, 165, 168, 171, 184, 185, 193–196, 199, 200, 213
Lander, F.W., 26
Lander's Cutoff, 26
Lander Technical Assistance Office, 196
land ethic, 187, 188, 207
language, 28, 76, 146, 202
lawyers, 182, 189
LeBeau, Tommy, 165, 170, 175, 178, 184
LeClair, Bud, Sr., 136, 214
Lemke, Tom, 77
Leopold, Aldo, 187, 188, 207
Lewis and Clark, 26, 35, 72, 83
Linkletter, Art, 152
Little Ice Age, 24
Little Wind River, 172
livestock grazing, 11, 195. *See also* cattle and sheep
logging, 65, 66, 187
Lolo National Forest, 41, 62, 79, 81
Lombardi, Vince, 209
Louv, Richard, 147

M

Manifest Destiny, 140
marmot, yellow–bellied, 153, 155
Marines, xii, 30, 63, 93, 94, 183
market hunting, 28
Martel, Wes (Wesley), 15, 17, 122, 199–205
McMullin, Steve, 208
Mesteth, Philip, 38, 40, 170, 214
methamphetamine, 200, 201
Mexican Pass, 130, 131
Michigan, ix, 4, 108, 184
Migratory Bird Treaty, 164
Miller, Jess, 125, 136, 137, 144, 204
Mississippi River, 10
Missoula, MT, 6, 81
Montana Cooperative Wildlife Research Unit, 6

Montana Fish, Wildlife and Parks Department, 77
Monument Peak, 99
moose: adaptations of, 87–88, 112; aerial surveys of, xii, 87, 134; behavior of, 88–91; evolution of, 35; food habits of, 38, 86, 88; habitat use, 41–42, 134; historical abundance, 134; hunting of, 134–135, 176–177; management, 185; occurrence on WRIR, 17, 37, 87; population estimates, 18, 70, 187, 197t
mortality of wildlife. *See* survival of wildlife
Moss, Pius, 125, 128, 184, 201, 204, 214
mothers, behavior with young, 60, 65, 78, 81, 84
mountain goats, 8, 20, 22, 59, 75, 79, 116
mountain lions, 116
Mountain Meadows, 101
Muir, John, 47
mule deer: food habits of, 72; habitat use of, 41–42; migration, 198; trophies, 32, 126; population of, 197t
Muskegon Lake, 43, 115
muskrat, 43, 44, 163

N

Nation, Ray, 38, 40, 46, 167, 168, 169, 170, 171, 194, 214
natural selection, 61
National Elk Refuge, 213, 215
National Outdoor Leadership School, 23
National Shooting Sports Foundation, 142
Nature, restoration of, 29, 66, 211
nature–deficit disorder, 147
Neal, Dan, 180
New York Times, 186
Nez Perce Indians, 121
Nipper, Art, Jr., 130, 131, 204, 214
Northern Arapaho. *See* Arapaho Indians

Noseep, Leon, 200
nutrition of young, 60

O

Oakleaf, Bob, 78, 215
Office of Aviation Services, 106
O'Gara, Bart, 6
oil, 11, 31, 107, 137, 163, 182, 187
old ways, 136, 148
O'Neal, Emil, 178–180, 182–184, 214
Orwell, George, 21
Outdoor Life Magazine, 189
Owl Creek Mountains, 35, 36, 38, 45, 46, 49, 62, 68, 70, 92, 96, 98, 107, 123, 125, 128, 129, 134, 160

P

Pacific Wagon Road, 26
Palm Desert, CA, 9
Palm Springs, CA, 9
parents of author, 146
Partners Against Meth, 200
Pengelly, Les, 122
petroglyphs, ix, 76, 133
Phillips, Bob, 4, 5, 7
Phlox Mountain, 130, 131
pika, 60
Plains Shoshones. *See* Shoshone Indians
politics, x, xi, 29, 31, 33, 209
Polson, MT, 92
Popo Agie River, 138
powwow, 164
prairie dogs, 72, 73, 180
predation, 35, 86, 136, 167
pronghorn: adaptations of, 35; aerial surveys, 102, 160; distribution of, 35, 160, 174; evolution of, 35; habitat of, 73, 172–173; historical population, x, 124–125; hunting of, 125, 135, 176, 190; importance to Indians, 124, 133; migration of, 68, 172, 174, 198; occurrence on WRIR, 14, 18, 115; population estimates, 197t, 198; study of, 51

Q

questionnaire, 36, 141

R

radiotelemetry (biotelemetry), 20, 106
Rancho Mirage, CA, 9
ravens, 173, 202
Red Creek, 71, 92
Red Desert, WY, 23
rehabilitation, wildlife, 208
Reisenbichler, Reg, 52–54, 214
religious beliefs, American Indian, x, 30, 164, 196, 204, 207
Religious Freedom Restoration Act, 196
reservations, Indian nations', x, 10, 12, 28, 139, 204
restoration, x, xi, 23, 29, 140, 141, 161, 198, 214
Riverside, CA, 4, 12
Riverton, WY, 23, 105, 134
Riverton Ranger, 180, 187
Riverton Reclamation Project, 135
Roberts, Father John, 120
Roberts Mountain, 56–58

S

Sacagawea, 120
sagebrush: as food, 37, 45, 71, 72, 172, 173; habitat, 20, 35, 40, 41, 73, 89, 138
sage–grouse, 5, 17, 51, 72
Sand Draw Highway, 173, 174
Santa Rosa Mountains, 5
Sapphire Mountains, 81
schools, 148, 150, 151
science: and culture, 202–203; education, 148–151; fields of, x, 12, 42–44; veterinary, 123; wildlife, xi, 145
Seattle, Suquamish Chief, 147
Selway–Bitterroot Wilderness Area, 116
Sequoia National Forest, 77
Shadow Mountain, 89
sharp–tailed grouse, 52

Shatto, Carl, 131, 214
sheep, bighorn. *See* bighorn sheep
sheep, domestic, 71, 131, 132, 133
Sheep Creek, 130, 131
Sheep Eaters (Shoshone), 37, 76
Sheridan, WY, 4, 51
Shoshone Episcopal Mission Boarding School, 120
Shoshone Indians, xi, xii, 13, 23, 26–31, 36, 37, 55, 62, 73, 115, 122, 141–144, 159, 180, 182, 185, 188, 190, 191, 203, 210, 211, 216: culture, 133, 144, 164, 178, 204; history, 14, 24, 26, 76, 124, 148
Shoshone National Forest, 71
Skates, Dave, 194, 196–198, 201, 215
skunk, 154
Smith, James, 121
Smith, Norma and Sandy (author's sisters), 42, 43
Smithsonian Institution, 28
snowfall, 68, 171
snowmobiles, 30, 117, 178, 186, 193
Snyder, Mrs. Charles, 164
Sonoran Desert, 46
species, threatened and endangered, 12, 58, 72, 150, 164, 176
species lists, 12, 17, 45–46
Standing Bear, Chief Luther, 1
Stagner, Martha, 131, 214
Stein, Herbert, 93
St. Stephens Indian High School, 149
substance abuse. *See* methamphetamine
Sun Bear, Chippewa, 193
Sun Dance, 164, 178
survival of wildlife, 20, 69, 72–73, 164, 171–173

T

tepee (lodge), 24, 120, 123, 138
Thermopolis, WY, 23, 93, 98, 101, 107, 118
Togwotee Pass, 193, 194
Trail Ridge, 96–98, 100, 108–110, 114, 118, 195

treaty rights (also vested or sovereign rights), 30, 125, 137, 187, 190, 191, 202, 209
Trout: cutthroat, 56; golden, 76, 77
Tubatulabal Indians, 76

U

umbrella species, 22
University of Montana, 14, 77
U.S. Constitution, 121
U.S. District Court, 196
U.S. Fish and Wildlife Service (USFWS), 4, 6, 7, 13, 46, 50, 59, 164, 190, 196
U.S. Forest Service, 7
U.S. Geological Survey, 75, 127
U.S. Supreme Court, 196
U.S. Wilderness Act, 52

V

vehicles, 4x4, 30, 136
Vietnam, 30, 102, 150
voting, tribal, 30, 149, 174, 175, 177, 178, 181–184, 190

W

wardens, game, 18, 66, 96, 127, 128, 135, 160, 172, 177, 189, 195, 214, 215
water: impoundments, 163; ponds and lakes, 52, 55, 65, 75, 76, 127, 146, 162, 163; rights, 13, 182
Ward, Alfred, 160
Warm Valley, 124
Washakie, Chief, 15, 24, 25, 26, 27, 29, 62, 120, 203
Washakie Shoshones. *See* Shoshone Indians
Washakie Wilderness, 128, 130
Watt, Interior Secretary James, 186, 188, 189
weasel, long-tailed, 165
Webber, Landis, 131, 214
Weed, Starr, 32, 33
Whiskey Mountain, 73
White Bear, Chief, 120
White-tailed deer, 36, 133, 197

Whitney, Mount, 76
wildlife management plan, WRIR, 7, 144–145, 163, 185, 187, 198
wildlifer (wildlife biologist), ix, x, xi, 6, 13, 18, 32, 33, 36, 50, 63, 84, 190, 194, 209
wildlife status reports, 159, 163, 185
willows, 27, 86, 87, 88, 96, 112, 113, 134, 162
Wilson, E. O., 44, 146, 209
Wind River, 33, 36, 52, 92, 107, 108, 129, 136, 162, 172
Wind River Agency, 46
Wind River Basin, 13, 35, 62, 124, 125, 134, 162, 187
Wind River Canyon, 34, 35, 128, 129, 130, 131, 134, 195, 210
Wind River Indian Reservation: area, xii, 7, 12, 14, 19, 27, 28, 29, 76, 99, 124, 125, 182, 185, 193, 199; history, 14, 122, 141, 144, 204, 214; population, 138, 200, 201, 208
Wind River Journal, 145, 164, 177, 178, 179, 187
Wind River Shoshones. *See* Shoshone Indians
Wind River Range ix, 35–38, 41, 44, 49, 50, 52, 56, 59, 62, 66, 68–71, 75, 76, 87, 92, 98, 99, 105, 112, 124, 127, 131, 132, 134, 135, 136, 160, 170, 186, 196
Wind River Roadless Area, 55, 187
wolf, gray, 20, 35, 83
Wounded Knee, 121
Wykee Lake, 75
Wyoming Department of Transportation, 173, 175
Wyoming Indian High School, 149, 181
Wyoming State Journal, 165
Wyoming State Veterinary Lab, 73

Y

Yellowstone National Park, 17, 23, 28, 66, 78, 83, 115, 127, 160, 164, 210